BEHAVIOR AND MIND

BEHAVIOR AND MIND

The Roots of Modern Psychology

Howard Rachlin

New York   Oxford
OXFORD UNIVERSITY PRESS
1994

Oxford University Press

Oxford   New York   Toronto
Delhi   Bombay   Calcutta   Madras   Karachi
Kuala Lumpur   Singapore   Hong Kong   Tokyo
Nairobi   Dar es Salaam   Cape Town
Melbourne   Auckland   Madrid

and associated companies in
Berlin   Ibadan

Copyright © 1994 by Oxford University Press, Inc.

Published by Oxford University Press, Inc.,
200 Madison Avenue, New York, New York 10016

Library of Congress Cataloging-in-Publication Data
Rachlin, Howard, 1935–
Behavior and mind : the roots of modern psychology / Howard Rachlin. p. cm.
Includes bibliographical references and index.
ISBN 0-19-507979-5
1. Philosophy of mind.   2. Psychology and philosophy.
3. Behaviorism.   4. Cognitive psychology.   I. Title.
BD418.3.R33 1994   128'.2—dc20
92-47398

1 3 5 7 9 8 6 4 2

Printed in the United States of America

Preface

There are two modern psychological sciences. One is cognitive or physi-ological psychology, which aims at the discovery of internal mechanisms, including complex mental mechanisms and representational systems. The other is teleological behaviorism, which aims at scientific explanation, prediction, and control of overt behavior, including the complex patterns of overt behavior that, I have come to believe, form our mental lives.

Behavioral analysis and the study of mental life seem like contrary if not contradictory activities not only in popular understanding but also in contemporary psychology and philosophy. This attitude, though under-standable, is wrong. Not only is behavioral analysis not really contrary to the study of the mind, behavioral analysis is potentially the best way to study mental *life* (as opposed to mental *mechanisms*) both of humans and other animals.

The misunderstanding of behaviorism in contemporary psychology is not just the behaviorists' concern; it is also a serious problem for psycholo-gy in general. Behavioral research, whether with rat, pigeon, monkey, or human subjects, is an essential link between academic psychology and vast areas of applied psychology, including clinical psychology, business, gov-ernment, and education. In other words, the thread of behavioral research and theory is actually a lifeline between psychological theories of *all* kinds and their real-world application.

In philosophy a corresponding misunderstanding obtains. Contem-porary philosophers of psychology have their internecine disputes but they are as one now in their contempt for behaviorism. Indeed, when one contemporary philosopher labels another a behaviorist it is either the prelude to or the conclusion of an attack—it is sometimes the whole attack. How could this state of things have come about?

As I began to read and attempt to understand the arguments of contemporary philosophers of psychology, I became convinced that their criticisms of behaviorism were misdirected. Philosophers were partly criticizing J. B. Watson's original molecular behaviorism (mental states identified as individual muscular movements) dating from early in this century and partly criticizing B. F. Skinner's rejection of mental terms in psychology. But mainstream philosophers did not and do not come to grips with the molar teleological behaviorism that forms the groundwork for my own empirical research.

J. R. Kantor (1963), a twentieth-century behaviorist philosopher far outside the mainstream, led me to the ancient Greek philosophers. Kantor claimed that modern behaviorism is an approximation (a poor approximation but the closest in modern times) to what he called "naturalism." Naturalism, according to Kantor, began in folk psychology, people's everyday attempts to deal with other people, and reached its peak with Plato and Aristotle. Aristotle's distinction between efficient and final causes as the two fundamental explanations of all movement forms the basis of the present book's distinction between cognitive or physiological psychology as a science of efficient causes and teleological behaviorism as a science of final causes. The discomfort of many modern philosophers with behaviorism stems, I believe, from a discomfort with final causes as explanatory principles. Therefore, one purpose of this book is *to justify final causes as scientific explanations of mental life*. Subsidiary to this purpose, the present book will argue that the sort of laboratory work that I and my colleagues have been doing, operant conditioning, was really all along the empirical arm of a teleological theory of the mind.

The cognitive psychologist George Miller titled his excellent introductory textbook *Psychology: The Science of Mental Life* (Miller, 1962). I argue here that the title of Miller's book might better have been used for a review of my own research and that of other teleological behaviorists. The actual contents of Miller's book would fit better under the rubric *Psychology: The Science of Mental Mechanisms*. The ultimate purpose of this book, then, its own final cause, is to answer this question: *How does the science of mental life (based on final causes) relate to the science of mental mechanisms (based on efficient causes)?*

Outline of the Argument

Loosely speaking, a psychology of efficient causes desires to answer the question: *How* does a given human or nonhuman being behave the way it behaves, feel the way it feels, think the way it thinks? The object of this sort of science is to achieve explanation, prediction, and control of the behavior, feelings, or thoughts of humans and nonhumans through causal connections among neural discharges, internal reflexes, internal mental representations, or other mental mechanisms. The philosopher J. A. Fodor (1981), in arguing for the central role of mental representations as internal efficient causes, says "Just for starters, try doing linguistics without recourse to mental representations" (p. 29). By implication all psychology involves mental representations. For Fodor, if an explanation of behavior fails to refer to mental representations, it is not a psychological explanation at all.

Despite the skepticism of Fodor and the majority of contemporary philosophers and psychologists, there exists another kind of modern psychology—a psychology of *final causes*, directed at the question: *Why* does this or that behavior, thought, or feeling come about? This nonstandard sort of psychology is explicitly teleological. Its fundamental principles were laid down in ancient times (in general terms by Plato and more specifically by Aristotle) and then forgotten in two stages; first, the ancient philosophers were interpreted in the West by scholars whose fundamental purpose was to provide intellectual grounding for Christianity; second, the methods of the ancients (as thus interpreted) were rejected by Renaissance physics. However the science of final causes as anticipated by Plato and codified by Aristotle is, I contend, particularly suited to psychology; *it was a mistake, at least in psychology, to have abandoned teleology.*

Psychologists have much to gain by studying Aristotle. Aristotle's psychology, according to one interpreter (Robinson, 1989, pp. ix–x) "is history's first fully integrated and systematic account and, to some extent, the one that remains the most complete." Although our current methodology of behavioral observation is more detailed, rigorous, and quantitatively sophisticated than Aristotle could have imagined, Aristotle's philosophical rationalizations and justifications for a teleological science are far more thorough and more thoughtful than modern psychologists have been able to devise; the breadth, the scope, the richness, the connectedness, the relevance to everyday life are greater in Aristotle's psychology than they are in modern psychology.

After discussing the modern psychologies of efficient and final causes and recalling how they developed, I shall explore in considerable detail Aristotle's own sciences of efficient and final causes. In this endeavor I have had to rely on modern translations and interpretations of various philosophical texts. The translations and interpretations I chose were those most congenial to my purpose and not necessarily the latest or the best classical scholarship. Plato, Aristotle, Augustine, and Descartes may really have meant something different from what I say they did, and the texts may very well differ from what I think they say. Whether my reconstruction makes sense in the context of their own interests, readers, even those who are not classicists, may judge for themselves. The purpose of these explorations therefore is not to argue with modern interpretations of classical philosophy, but rather to draw inspiration from them for modern psychology.

For Aristotle, the question *why* necessarily precedes the question *how* in order of investigation. Whether or not we accept that point, we all may agree that both questions are important and that answers to one may throw light on attempts to answer the other. Cognitive psychologists have recently begun to express dissatisfaction with the fractionation, approaching chaos, in modern cognitive theory (Loftus, 1985; Watkins, 1990). Knowledge of the final causes of behavior might perhaps aid in the attempt to discover efficient causes thus giving direction to modern cognitive psychology. Consider again the quote from Fodor: "Just for starters, try doing linguistics without recourse to mental representations." Linguistics and much of psycholinguistics ask: How do we say what we say? Perhaps Fodor is correct that mental representations are essential in answering this question. But the psycholinguist may be interested also in *why* we say what we say and not just incidentally interested; our goals may well influence our methods and our goals may have nothing to do with mental representations.

Aristotle claimed that good physicists (psychology for him being a branch of physics) need to know all the causes of their object of study. In the case of behavioral dynamics these include final causes as well as efficient causes—why as well as how. Modern academic psychology is focused almost exclusively on efficient causes. Until the science of final causes in psychology has developed sufficiently to become part of the standard curriculum, perhaps this volume will serve as a handbook for those with faith that it will do so.

One further point: I have tried throughout to illustrate difficult arguments and issues with examples from everyday life. When the examples are my own I have chosen them from modern life so that the reader may clearly see that they are mine rather than those of an ancient philosopher. A discussion of Aristotle's concept of sensation in terms of a driver's discrimination between red and green lights is thus obviously my illustration rather than Aristotle's.

New York   H. R.
February 1993

Acknowledgments

I would like to thank William Baum, Hugh Lacey, Albert Silverstein, and Gerald Zuriff for their perceptive criticisms of an earlier draft. Whatever organization and coherence the book contains is in large part due to them. I would also like to thank my teacher, Richard Herrnstein, whose articles (especially, 1969, 1970) have given form to teleological behaviorism.

Preparation of this book and the research described herein was supported by grants from the National Institute of Mental Health, the National Science Foundation, and a residency at the Russell Sage Foundation. Anthony Castrogiovanni and Eric Siegel, two of my students, aided greatly in the conception and drawing of the figures. I owe a special debt to Barbara Lento whose skill and care in preparation of the manuscript were very much needed.

Contents

BEHAVIOR AND MIND

1

Behavior, Cognition, and Mind

Behaviorism is a term for a set of ideas more or less prevalent in American psychology between 1913 and the present. In 1913 John B. Watson (1878–1958) published an article, "Psychology as the Behaviorist Views It," which defined behaviorism as a method and proposed it as the only valid method for psychology. Since then, many American psychologists have called themselves behaviorists. Most prominent among them have been Clark L. Hull (1884–1952), Edward C. Tolman (1886–1959), and B. F. Skinner (1904–1990). There have existed at various times, more or fewer followers of these theorists—Watsonians, Hullians, Tolmanians, and Skinnerians—but the behaviorists themselves never formed a single school in the way that Freudians initially did. (See Zuriff, 1984, for a comprehensive analysis of behavioristic thought.)

At the very beginning with Watson, and at present, when they have been under considerable attack from nonbehavioristic psychologists and philosophers, behaviorists have attempted to examine and justify their beliefs, in the course of which they have, quite naturally, spent much of their energy attacking each other. These arguments have given rise to great diversity among their views. Nevertheless there are some outstanding commonalities among them in theory and practice. First, they all reject introspection as a method for discovering psychological truth. The rejection of introspection was Watson's main methodological point; it has never been abandoned by behaviorists, and the point has never been strongly challenged by American psychologists of any school.[1]

The Rejection of Introspection

Most behaviorists, including all of the prominent ones mentioned above, have experimented, most of the time, with nonhuman animal subjects. Watson has declared that the practice of experimenting with animals led directly to his rejection of introspection. For other behaviorists, it is not clear which came first: principle or practice. Most likely both came together; the resolution not to accept introspective reports as valid indicators of mental events is easily followed when there are no such reports among the data and the lack of such reports among the data leads one to declare that they are unnecessary in theory.

Behaviorism arose in the United States about the same time as gestalt psychology in Europe. Both were reactions to the psychology of Wilhelm Wundt (1832–1920), usually considered the founder of experimental psychology (Boring, 1950). Wundt believed that when people introspectively report their own sensations, they are describing physiological events. Introspection for Wundt was a form of physiological psychology. Whereas normal physiology sees the nervous system from the outside, introspection sees it from the inside; just as vision can be directed to various parts of an external object, so introspection can be directed toward various parts of the interior of the body; just as with training a person can learn to make fine discriminations among visual stimuli, so with training a person can learn to make fine discriminations among internal events. The power attributed to this process was such that university authorities at Leipzig (where Wundt had established the first psychological laboratory) hesitated to allow students to be trained in introspection lest they be driven insane (Boring, 1950).

Wundt was trained as a physician and physiologist. At first he was unsuccessful but, according to Fancher (1990, p. 153), "Wundt's fortunes improved markedly after 1874, when he . . . completed the two volumes of *Principles of Physiological Psychology.* In this landmark book Wundt not only defined a 'new domain of science' whose task was to conjoin the two previously separate disciplines of physiology and psychology, but also provided detailed examples of how the task could be accomplished. In providing the first genuine textbook for the new field, Wundt emphatically established himself as its leader."

Wundt's teacher of physiology at Berlin was the preeminent physiologist of his day, Johannes Müller (1801–1858). To understand the role of introspection in Wundt's psychology, it is necessary to consider Müller's own psychology, which is expressed in his famous *law of specific nerve energies.* According to E. G. Boring (1950, p. 80), the "doctrine of specific energies of nerves, as Johannes Müller named it, was the most important law in sense-physiology which these early decades [of the nineteenth century] produced. The law is associated especially with Müller's name because he had the most to say about it and insisted most emphatically upon it. . . ." Boring (1950, p. 82) characterizes the law as follows:

The central and fundamental principle of the doctrine is that we are directly aware, not of objects, but of our nerves themselves; that is to say, the nerves are intermediates between perceived objects and the mind and thus impose their own characteristics upon the mind.

Müller's dictum was: "Sensation consists in the sensorium's receiving through the medium of the nerves, and as a result of the action of an external cause, a knowledge of certain qualities or conditions, not of external bodies, but of the nerves themselves."

The mind was thus seen by Müller as a prisoner in the body, separated from the world by the nerves. According to Müller the mind's function is to overcome the limitations of that isolation by reconstructing the world out of the incoming sensory information. Müller was therefore not far from modern cognitive psychology. Although there were no computers in Müller's time, the concepts of information processing and mental representation were present in his theory. The trichromatic theory of color vision and the basilar-membrane place theory of hearing developed by Hermann von Helmholtz (Müller's earlier student to whom Wundt at one point served as an assistant) may be seen as theories of information processing. Some modern cognitive theories (e.g., Shepard & Metzler, 1971) are explicitly of this kind, and the main method used to test them (subtraction of reaction times) is taken straight from Helmholtz and Wundt (although direct introspection of internal processes has been generally avoided by modern cognitive psychology). Müller's conception of the mind was, in its basic form, passive. That is, the information comes in from the senses, and the mind, according to its innate dispositions, combines it with prior input and processes it automatically. Wundt's crucial contribution was the notion of mental *activity,* that at least parts of this process could be observed *from the inside* by an action of *will.* Wundt himself was cautious about the power of introspection to reveal higher mental processes (beyond sensations and feelings). However, his student, Edward Bradford Titchener (1867–1927), was much less cautious. Titchener's views are important because after receiving his doctorate in 1892 (Wundt was his adviser) he established at Cornell University the largest and most influential psychology laboratory in the United States. It was Titchener's version of Wundt's psychology to which Watson's behaviorism reacted.

Titchener (following his British compatriot John Stuart Mill [1806–1873]) viewed the mind as composed of sensory elements structured much like chemical elements in a compound. The elements were supposed to consist of the remnants of past sensations (colors, tones, smells, and so forth) having only a few dimensions (quality, intensity, duration). In Titchener's psychology, as in Wundt's, overt behavior played no important part. While Wundt and Titchener might have argued with John Locke (1632–1704) about the essential physiological substrate of the mind and the innateness of mental contents, the image conceived by Etienne Condillac ([1715–1780] a French follower of Locke) of an immovable statue with a mind inside of it (capable of processing information) would have been no less conceivable to Wundt and Titchener than to Locke and Condillac.

For Wundt and Titchener mental activity (such as attention, reasoning, judgment), like mental stimulation (sensations, feelings), was wholly *internal*. In that sense these psychologists reflected European philosophy of the time, both on the Continent and in Great Britain.[2] Behaviorism was a step away from that philosophy, directed by the pragmatism of William James (1842–1910) and John Dewey (1859–1952).

Watson objected first to the methods of Titchenerian structuralism, which stressed introspection, but later Watson and other behaviorists began to object to the very purposes of structuralism. Wundt's and Titchener's methods were designed to get *past* overt behavior, which was seen only as a sign or messenger or ambassador of a mental world existing inside the human or nonhuman subject. Those methods would not be expected to retain their efficacy if the object of psychological study were to predict, control, explain, or find patterns in behavior itself. In the hands of the behaviorists, introspection lost its primacy as a window into the mind and became just another bit of behavior to be fit into a pattern with the rest.

Gestalt Psychology

Gestalt psychology, an outgrowth of early twentieth century German philosophy, was a much different sort of objection to Wundtian structuralism than was behaviorism. The gestalt psychologists objected, not to introspection, but to the way in which it was used by Wundt. For Wundt, psychology, in order to discover the naive contents of consciousness, had not only to get past the outside world, past behavior, but also past the accretions that experience had added to introspection. Introspectionists had to be trained not to make "the stimulus error"; that is, not to report as introspections what were actually conclusions about the nature of the world. Naive introspections, according to Wundt and Titchener, consisted of simple sensations—brightnesses, loudnesses, odors, etc.—and their combinations. The gestalt psychologists differed from Wundt and Titchener on what were fundamental contents of consciousness. For them, the fundamental contents were perceptions—not brightnesses or loudnesses but books, trees, people—forms and objects of the world, distorted perhaps, but topologically intact. Wundt's conception of the elements of consciousness was, according to the gestaltists, the product of another kind of error, called "the experience error," that came from trying to break the mind into elements as if it were a chemical compound. Such analysis, they argued, is inappropriate for mental events, which are governed by laws much more like those that govern physical fields.

Gestalt psychology was brought to the United States by Max

Wertheimer (1880–1943), Wolfgang Köhler (1887–1968), and Kurt Koffka (1887–1941) who emigrated here in the late 1930s. Their ideas, especially as they criticized Wundtian structuralism, came to influence the thinking of the American behaviorists. This influence was explicit in the case of Tolman. "Sign-gestalt expectancies," for instance, were for Tolman a fundamental kind of learning: what to expect from a situation or a "behavioral field" on the basis of exposure to a portion of that field. For instance, upon opening a door to a classroom, you expect to find, on the other side, a floor beneath your feet and a blackboard, desks, and chairs arranged in a certain general way different from the arrangement of objects in an office. There may be no key element that differentiates a classroom from an office. It is the form (*gestalt*) that is different and this difference could be learned.

Although the gestaltists felt consciousness to be both scientifically accessible and relevant to overt behavior, they denied the existence of mental processes as distinct from physical ones; they believed in what they called *isomorphism* between mind and brain. While Wundt did see the mental and physical as two aspects of the same thing, he believed that experiences of "will" and "voluntary effort" (the forces that guide introspection to various aspects of the inner world) as well as "creative synthesis" were explicable only in terms of "*psychic causality* whose rules are not reducible to the purely mechanistic processes of physical causality" (Fancher, 1990, p. 164). But "psychic" causes were not, for Wundt, final causes (more's the pity). They were rather, Wundt implied, efficient causes emanating from an internal immaterial entity such as Descartes believed was the origin of voluntary behavior.[3]

In rejecting Wundt's dualism, the gestalt psychologists believed that they came closer to contemporary behaviorists than did the structuralists. This was true as far as it went. However, their model of the mind was much closer to modern materialism (what I later call *physiologism*) than it was to behaviorism. Skinner (1979, p. 246) quotes his own contemporary description of a lunch with Köhler and Kurt Lewin (another prominent gestaltist). Skinner says: "We had a hell of a violent argument. I don't know what they were trying to do—convert me I guess. Strangely enough we finally located our difference pretty clearly, and that was that."

The lasting contribution of the gestalt psychologists to American psychology was their focus on context—their insistence that nothing is perceived or learned in isolation. Everyone knows that a gray spot on a white background appears black while the same gray spot on a black background appears white. The gestalt psychologists claimed, further, that the two spots differ in brightness not because they are judged differently and not because the different backgrounds fool the observer, creating an illusion (that might be overcome by deeper introspection) but because of a fundamental fact of phenomenal experience; what you perceive in the first place, the gestaltists claimed, is the *relationship* between the spot and its background. So, at the most fundamental level of perception, the two gray spots differ in brightness. Context is thus not a correction to perception or a distorter of perception but a vital component of the fundamental perceptual process.

The gestalt psychologists demonstrated over and over again that context is important, not only in perception but also in learning, the area where behavioral theory was most directly applied. Animals seemed able to learn relationships directly. A chicken may apparently learn to peck at the larger of two circles, rather than at a circle of a particular diameter. A monkey seems able to see the functional relationship between a stick in its cage and a banana out of reach beyond the bars. The problem such demonstrations caused for the behaviorists was that while the specific actions of a person may be easily observable, the larger context of those actions (what they are relative to) often seems hidden within the person's mind and discoverable only by introspection. We now turn to a discussion of how behaviorists attempted to cope with this problem.

Context

Although behaviorists agreed with each other that introspection was an unreliable psychological method, they have never been able to agree on a method to replace it. Introspection serves an important function, even from a behavioristic point of view; it provides a context for immediate behavior. Suppose a boy is offered an ice cream cone and takes it. So much is overt behavior. But the boy may be taking the ice cream cone because he wants to eat it or he may just be taking it out of politeness. One way to tell the difference is to ask him. But if you accept his answer as a valid indication of his motivational state, you are accepting the validity of his introspections, for his answer appears to come from nowhere else.

Or, consider an event in a psychology experiment where a woman chooses a red card with a green triangle on it in preference to a yellow card with a blue circle on it. Did she choose on the basis of color or shape or position, or a combination of these factors, or some other factor unknown to the experimenter? Obviously, you could ask her. But you then are assuming that, in answering your question, she refers to some internal state or process unavailable to you. If you believe that introspection is a window into the mind, such an assumption has face validity. When the woman says she was choosing on the basis of color, she gives you critical information on the essential context of her immediate behavior. But if you assume that introspection is just more behavior, her words give you only more pieces of behavior to analyze, pieces that may be quite irrelevant to what you consider to be the critical reasons for her original choice—the essential context, the causal factors.

It is on the nature of those critical causal factors that behaviorists differ. More generally, they differ on how to use the language that has evolved among humans for describing behavioral context, the language of mental terms: hopes, beliefs, wishes, thoughts, feelings, etc. If these terms are not descriptions of the contents of consciousness, available to intro-spection, then to what do they refer? What do mental terms mean?

Watson, Hull, and Tolman

For Watson, the answer was provided by Ivan Petrovich Pavlov (1849–1936), the great Russian reflexologist. Mental terms, according to Pavlov, described reflexes of the central nervous system. Thoughts were muscular movements or, if not actual muscular movements, then unobserved central nervous system events. Complexity of the mind was nothing but complex-ity of the reflexes. The context of immediate behavior was, for Watson, a physiological context (physiology as studied by reflexologists using mostly nonhuman subjects—not the particularly human interior vision in the physiology of Wundt).

Hull, at first, provided an account of the context of immediate behav-ior in terms of conditions of deprivation and incentive and the animal's previous exposures to the current task. He hoped that the goal-directed behavior of animals, as exemplified by the behavior of rats in mazes, could be fully explained without the use of mental terms; then whatever princi-ples were discovered to account for the behavior of rats could be carefully extended to apply to the behavior of humans. Terms to describe the con-text of immediate behavior would then be extensions of the terms used to describe the immediate behavior itself. The rat was to be for psychology what the vacuum was for physics, a sphere of action where the laws of behavior operate simply. But, mostly due to the work of Tolman and his students, it became obvious that the immediate behavior of even so simple an animal as the rat could not be explained without further contextual extensions. For instance, it had been shown (evidence summarized by Kimble, 1961) that if a rat, neither hungry nor thirsty, had been allowed to explore a maze with food and water in different spots, the rat, later de-prived of water, would run right to the spot where the water was or, later deprived of food, would run right to the spot where the food was. The rat seems to have *known* where the food and water were.

In response to Tolman, Hull and his followers adopted a solution much like Watson's. Corresponding to the external stimuli and overt re-sponses that had been defined as functional entities, Hull (1952) postu-lated internal connections between stimuli and responses to serve as refer-ents for talk about mental states of the rat. These internal connections (symbolized as r_g–s_g's), like the internal reflexes of Watson and Pavlov, became scientific-sounding substitutes for thoughts and emotions.

Because r_g's and s_g's were supposed to be functionally identical to observable responses and stimuli, their properties were in principle link-able to physical data language. However, Dewey (1896) had warned American psychologists that the concepts of stimulus and response have no *psychological* meaning in the interior of a behaving animal. Dewey had argued that, because physiological systems are complex networks full of internal feedback loops, each affecting most of the others, a chain of stimuli and responses could not even in principle be traced through an animal.

It was obvious to the gestalt psychologists and to Tolman, their main supporter among American behaviorists, that Watson's and Hull's solutions to the problems of context of behavior contained all of the disadvantages of Wundt's and Titchener's original structuralism. Just as Titchener claimed that any introspective report was reducible to individual sensations and their connections, so did Watson and Hull claim that any behavior was reducible to individual reflexes. Tolman saw the behaviorisms of Watson and Hull as simply motorized versions of the structuralism of Titchener.

Tolman, calling himself a "molar behaviorist," began in the 1930s and 1940s to attempt to restore mentalistic vocabulary to descriptions of the behavior of animals, particularly rats learning to negotiate mazes. The rats were said to develop hopes, fears, and expectancies for rewards and cognitive maps of the mazes. Such terms had been applied to the behavior of animals by the early (pre-Watsonian) functionalists [Conwy Lloyd Morgan (1852–1936) and George J. Romanes (1848–1894), for example] on the basis of the introspections of the observer. The method of the early functionalists was first to observe an animal's behavior and then to introspect on what mental operations would be minimally necessary to perform it. Lloyd Morgan's "canon" demanded that the least advanced or "lowest" mental operation conceivable (by the observer) to have produced the observed behavior be ascribed to the subject. The very first study of a rat in a maze reported in 1901 by Willard Stanton Small (1870–1943) a functionalist at the University of Chicago, used this method. Following is an excerpt from Small's report of the results (from Herrnstein & Boring, 1965, pp. 552–553):

> In appreciating the results of this series of experiments . . . the [following] . . . facts come to view . . . the initial indefiniteness of movement and the fortuitousness of success: the just observable profit from the first experiences; the gradually increasing certainty of knowledge indicated by increase of speed and definiteness, and the recognition of critical points indicated by hesitation and indecision: the lack of imitation and the improbability of following by scent: the outbreak of the instincts of play and curiosity after the edge of appetite is dulled. In addition are to be noted the further observations upon the contrast between the slow and cautious entrance into, and the rapid exit from the blind alleys, after the first few trials; the appearance of disgust on reaching the end of a blind alley; the clear indication of centrally excited sensation (images) of some kind; memory (as I have used the term); the persistence of certain errors; and the almost automatic character of the movements in the later experiments. Viewed objectively, these observations all converge towards one central consideration; the continuous and rapid improvement of the rats in threading the maze, amounting to almost perfect accuracy in the last experiments. No qualification of this view was found necessary in the light of many later experiments. Rather they all confirm it.

The mental aspect is considerably more complex, the mental factors, much more difficult of analysis and evaluation; but the central fact in the process seems to be in the recognition by the rats of particular parts of the maze.

Tolman hoped to restore the use of mentalistic language in descriptions of animal behavior, but without the step requiring the observer to introspect. Tolman attempted to specify objective criteria for mental phenomena such as expectancies. If the experimenter were to say that a rat had an expectancy, the rat would have to pass certain strictly behavioral tests. For instance, to show that a rat expected reward A, the rat's pattern of behavior would have to be disrupted (the rat would have to run through the maze significantly slower or faster) after reward B was substituted in the goal box for reward A. Once these tests were established, the observer's job would be simplified to activities that presumably required no introspection at all, such as readings of counters and timers. Had Tolman's program fully succeeded, it would have been possible (given modern computer technology) to build a machine, the output of which would be a slip of paper saying: "The rat in this machine, at time T, expected food to be in the goal box."

Tolman's rejection of introspection underlay his claim to be a behaviorist. To those who argued that the initial setting of the criteria for the ascription of mental terms to animals implied prior introspective activity on the part of the observer, Tolman could say that such introspections were temporary and unimportant. Once adequate behavioral criteria for mental activity were established, the introspections on which they might have been based could be dispensed with. For Tolman, as a behaviorist, the ultimate test for the adequacy of a given set of criteria for ascribing expectancy to a rat was not whether those criteria conformed well or poorly to the observer's introspections but whether they served well or poorly in predicting and controlling the rat's behavior.

Like Hull, Tolman believed that the rat in the maze could serve for psychology as an arena for the observation of simple behavioral laws. Tolman differed from Hull in his use of mental terms to express those laws. The debate between Tolman and Hull corresponds to debates that have occurred and are occurring about the use of mental terms in biology. For instance, an almost identical debate occurred at the turn of the century between the biologists Herbert S. Jennings (1868–1947) and (Watson's teacher) Jacques Loeb (1859–1924; see Loeb, 1918). Jennings took the position that mental terms could be useful to describe the behavior of even one-celled organisms *provided* objective criteria were given for the use of those terms. According to Jennings (1906; from Herrnstein and Boring, 1965, p. 481):

We do not usually attribute consciousness to a stone, because this would not assist us in understanding or controlling the behavior of the stone. Practically indeed it would lead us much astray in dealing with such an object. On the other hand, we usually do attribute consciousness to the dog, because this is useful; it enables us practically to appreciate, foresee, and control its actions much more readily than we could otherwise do so. If Amoeba were so large as to come within our everyday ken, I believe it beyond question that we should find similar attributions to it of certain states of consciousness a practical assistance in foreseeing and controlling its behavior. Amoeba is a beast of prey, and gives the impression of being controlled by the same elemental impulses as higher beasts of prey. If it were as large as a whale, it is quite conceivable that occasions might arise when the attribution to it of the elemental states of consciousness might save the unsophisticated human being from the destruction that would result from the lack of such attribution. In such a case, then, the attribution of consciousness would be satisfactory and useful. In a small way this is still true for the investigator who wishes to appreciate and predict the behavior of Amoeba under his microscope.[4]

Loeb claimed on the other hand that the behavior of all animals, even human beings, could and should be explained without the use of mental terms. Loeb invented another language, that of tropisms and associative memory (as a purely physical process), which he felt more adequately, and scientifically, classified behavior.

While Tolman insisted that evidence for mentality in animals be behavioral evidence he still maintained (at least in his later works) that the behavior was *evidence;* that is, that something existed other than behavior that the behavior was evidence *of.* An expectancy or a cognitive map for Tolman was something other than the observed pattern of the behavior. In his most influential work, *Purposive Behavior* (1932, p. 428), Tolman wrote: "It is clear that . . . means-end readinesses and expectations are logically and usually also temporally, prior to the realities [the behavioral observations] which would verify them." Although Tolman was an ontological monist (rejecting the transcendence of mental states), he was an epistemological dualist. People's knowledge of their own mental states is for Tolman *in principle* different from an observer's knowledge of their mental states. Behavior was for Tolman (as it is for many cognitive psychologists) merely evidence from which to infer cognition—much as symptoms for a physician are evidence from which to infer a disease. An expectancy or cognitive map, as Tolman conceived it, was not the behavior itself but an intervening variable mediating between a (globally defined) stimulus and a pattern of behavior.[5]

Skinner

Like Watson (1913), Hull (1943), and Tolman (1932), Skinner (1938) attempted to find, in the laboratory, a condition where the laws of behavior act simply (a behavioral vacuum) but, unlike the other three behaviorists. Skinner did not look for simplicity in the nervous system of an animal. Instead, he sought simplicity at the boundary between the animal and its environment where contingencies of reinforcement are said to act. The Skinner Box was designed to eliminate all factors irrelevant to those contingencies, to be a simple environment where simple laws of behavior would emerge. In such an environment, Skinner felt, it should not matter if the animal is a rat, a pigeon, a monkey, or a person; once the ways in which different species make contact with the environment have been taken into account, the contingencies should shape the behavior in the same way.

Skinner's behaviorism is more extreme than were Watson's, Hull's, or Tolman's. Like Watson and Hull, and contrary to Tolman, Skinner rejected the use of mental terms in behavioral description. But Skinner's solution to the problem of context (how to describe the difference between a boy who takes an ice cream cone because he wants to eat it and a boy who takes an ice cream cone to be polite, or between a subject in an experiment who chooses a red triangle because it is red and one who chooses a red triangle because it is a triangle) is different from that of Watson and Hull. Watson, Hull, and Tolman held that the context of immediate behavior lies inside the organism. For Watson, the boy who wants the ice cream cone and the boy who is just being polite differ in the states of their internal reflexes; for Hull, they differ in their internal fractional anticipatory goal responses $(r_g-s_g's)$; for Tolman, they differ in their internal cognitions. But for Skinner, the two boys differ in their "reinforcement histories," which lie not inside the boys, but in the interactions that have occurred between the boys and their environments. Reference to the difference in reinforcement histories to which the two boys have been exposed is, according to Skinner, both necessary and sufficient to describe the difference between them. How do you find out these histories? In the complexity of everyday life, an act's reinforcement history is no more accessible than its underlying neural activity. A reinforcement history (much like a political or social history) is a *theory*. But, Skinner argued, just as a physiologist might find evidence for or against one or another neural theory in the physiological laboratory, so the psychologist might find evidence for or against one or another reinforcement-history theory in the operant laboratory—where reinforcement history may be, to some degree, controlled.[6]

Most important, a reinforcement history may best be described without the use of expressions such as "wanting to eat the ice cream cone" and "wanting to be polite." The language of reinforcement history refers to contingencies of reinforcement and rates of response under those contingencies. To Skinner, the difference between the two boys is in principle like the difference between two rats, both of which, at a given instant, are pressing a lever (L) but one of which has been rewarded with food (F) contingent on one pattern of responses, A-B-C-L-D-F and the other of which has been rewarded with water (W) contingent on another pattern of responses, G-H-I-J-L-W. The relevant context of the lever press of each rat is not anything that occurs inside the rat but is the other overt behaviors that the rat exhibits, together with the contingencies of reinforcement to which it has been exposed. Thus, two boys who reach out for ice cream cones differ in other acts that form part of a behavioral pattern. For the first boy, the pattern of behavior that forms the context of reaching out for the ice cream cone may be in the same category as the pattern that forms the context of reaching out for cookies and reaching out for candy bars. For the other boy the pattern of behavior in reaching out for the ice cream cone may be in the same category as that involved in saying "please" and "thank you," and helping old ladies to cross the street. Or, the pattern may be in both categories for both boys but the relative dominance of the categories may differ. Reinforcement history (along with genetic history) defines the categories and sets their relative dominance.

For Skinner, reinforcement is the ultimate shaper of behavior. Behavioral categories are partly determined by the biology of the animal and partly by the environment. The categories are called "operants" by Skinner. The action of reinforcement on operants is called *operant conditioning*. Skinner (1974) believed it would be misleading to use mentalistic language in operant conditioning. Mental terms can refer meaningfully only to an animal's reinforcement history yet they are almost always interpreted as referring to internal states. Thus it is better not to use them at all in any attempt to treat behavior scientifically; it is better to keep to a language that was developed to deal with reinforcement history as such.

Teleological Behaviorism

Skinner's recent books and articles have been attempts to show the necessity and sufficiency of the language of operant conditioning in situations in which mentalistic language is typically used. Some followers of Skinner, post-Skinnerian, teleological behaviorists, have begun to use mentalistic language in the analysis of behavior of animals in a way reminiscent of Tolman's use of mentalistic terms in the face of Hull's theories of learning. Teleological behaviorists view mental terms as descriptions of molar behavior—acts extended in time.[7] This view is like Tolman's in the sense that temporally separated environmental events are used to define mental terms but like Skinner's in the sense that environmental events themselves, not internal events, are the subjects of interest.

Teleological behaviorists see overt events and the behavior related to them as constituting the meaning of mental terms. This differs from Skinner's more molecular view, from which mental terms are seen as misleading names for specific acts. I shall elaborate upon this difference in Chapter 2 but for now I summarize it by saying that teleological behaviorists share *molarity* with Tolman and *externality* with Skinner. My expectation of a birthday present, for example, might have been described by Tolman as a behavioral event; for example, the sentence, "Gee, I can't wait for tomorrow," uttered on the day before my birthday *plus* its context. Tolman's view is *molar* because for him both the event and the context would be equally important components of my expectation. But for Tolman the context of the particular act would be an internal state—a mental representation.

Teleological behaviorists *externalize* the context as well as the act. Thus, for a teleological behaviorist the relevant context of my words on the day before my birthday might consist of semantically similar sentences uttered on this and other days-before-birthdays and birthday presents received.

Five Ways of Using Mental Terms

The brief discussion of how mental terms have been used in psychology was meant to clarify the currently vague distinction between mentalists, cognitivists, physiologists, and behaviorists. Any attempt to characterize these viewpoints must be arbitrary. But it will be impossible to go further without some attempt to define them. Accordingly, the following quite arbitrary definitions are set forth below. The object of the definitions is to distinguish each viewpoint from the others, not necessarily to reflect the opinions of any theorist or group of theorists.

Mentalism: The belief that mental terms refer to internal intrinsically private events that may be revealed by introspection.

Physiologism: The belief that mental terms refer most precisely to internal events that occur in an animal's nervous system. These events are revealed, not by introspection, but by physiological investigation.

Cognitivism: The belief that mental terms refer to internal events, reliably revealed neither by introspection nor by physiological investigation but by overt behavior including verbal behavior. From a careful analysis of overt behavior it is possible to infer the existence of internal events as one might infer the program of a computer from its inputs and outputs. Just as a given computer program may be instantiated in hardware in any number of ways (a computer memory with a given function in the program may consist of bubbles, transistors, vacuum tubes, relays, or other devices), cognitive psychology as such is not committed to any particular physiology.[8]

Skinnerian behaviorism: The belief that all of the behavior of animals—including humans—may be explained in terms of prior stimulation (the cause of "involuntary" behavior) and contingencies of reinforcement (the cause of "voluntary" behavior). All behavior usually considered to be caused by the mind may be reinterpreted in terms of the animal's reinforcement history and natural selection of its species. Mental terms are therefore not properly a part of psychology.

Teleological behaviorism: The belief that mental terms refer to overt behavior of intact animals. Mental events are not supposed to occur inside the animal at all. Overt behavior does not just *reveal* the mind; it *is* the mind. Each mental term stands for a pattern of overt behavior. This includes such mental terms as 'sensation,' 'pain,' 'love,' 'hunger,' and 'fear' (terms considered by the mentalist to be "raw feels"), as well as terms such as 'belief' and 'intelligence' that are sometimes said to refer to "complex mental states," sometimes to "propositional attitudes" and sometimes to "intentional acts."

Teleological behaviorism is intuitively hard to accept. Our very language goes against it. Chapter 2 will attempt to further define this mode of thought, and to discuss its implications. Although no behavioristic argument can rest merely on appeals to intuition, the following personal anecdote may give the reader at least a preliminary sense of the direction taken by the succeeding chapters.

Several years ago I took part in a debate before an audience of philosophers at the New School for Social Research in New York City. I argued that pain, for all its undeniable horribleness, is really a public event and not, as most philosophers believe, a private internal "raw feel." (This argument about pain will be presented briefly in Chapter 7.) I implied further that pain was perhaps the most prototypical example of a mental state that seems to be indisputably inside us. But, like all mental states, pain, I argued, may be understood as a pattern of overt actions, better observed by someone else than by the person having the experience.

After the lecture a prominent philosopher in the audience (a former teacher of mine) raised his hand and asked me if I believed that love was, like pain, "just" a pattern of behavior. Yes, I said; pain, love, consciousness, intelligence, our very souls, are nothing but patterns of overt behavior. But then, the philosopher said, my argument was easily refuted by the "mechanical dolly example."

He asked me to suppose that I were single and one day met the woman of my dreams—beautiful, brilliant, witty, and totally infatuated with me. She asks me for a date, we go out and have a wonderful evening. We spend the night together and, the philosopher implied, I have the best sexual experience of my life. The next morning, however, she reveals to me that she is not a real human being but a mechanical doll, composed of transistors, circuit boards, silicon, and steel. Wouldn't I be disappointed? I admitted that I would be disappointed. Doesn't that prove, the philosopher said, that what I valued in the doll was not her external appearance and external behavior, for these were by hypothesis as good as they could be, but her internal qualities, her internal soul, of which her external appearance and external behavior were only signs? I would be disappointed, the philosopher implied, because I would have gotten only a few signs of love but not real love, which is truly a feeling of the heart, an emotion, something only a real human being can have. The mechanical doll, because she was just going through the motions without having the feeling, could not possibly love me—hence my disappointment.

This argument does have some force, and at the time I did not have the presence of mind to answer my old teacher in any adequate way. But now, with the benefit of hindsight, let me tell you what I should have said.

I should have asked the philosopher to imagine another mechanical doll (Dolly II, an improved model). This doll, as beautiful, as witty, as charming, as sexually satisfying as Dolly I, doesn't reveal to me, the next morning, that she's a machine; she keeps it a secret. We go out together for a month and then we get married. We have two lovely children (half-doll, half-human) and live a perfectly happy life together. Dolly II never slips up. She never acts in any way but as a loving human being, aging to all external appearances as real human beings do—but gracefully, retaining her beauty, her wit, her charm. Finally, inevitably, on the exact date of our fiftieth wedding anniversary, after the celebration with our loving children, grandchildren, and great-grandchildren, she calls me to her and tells me that she is about to die (rust has taken its toll). She has one last request. "Please," she says, "don't let them do an autopsy." But in my grief I forget. She dies, an autopsy is done, and it is *then* revealed to me that my doll was not a "real" human being but only composed of transistors, silicon, and steel. Would this knowledge lessen my grief? I have spent 50 years married to Dolly II, every day better than the one before. And I have lost her forever. Would the knowledge that the chemistry of her insides was inorganic rather than organic make any difference to me or her loving children or her friends? I don't think so.

What then is the critical difference between Dolly I and Dolly II that causes me to be disappointed with one and just the opposite of disappointed with the other? Obviously the difference cannot fundamentally depend on what goes on inside them. The difference (to use an Aristotelian metaphor) is that Dolly I is like a single swallow and Dolly II is like a whole summer. As my old teacher implied, Dolly I's appearance and behavior was just a sign of something that wasn't really there, was "mere behavior" in the absence of a human soul. But the story of Dolly II reveals that the thing that wasn't there in Dolly I—the soul, the true love—consists of *more* behavior. One swallow does not make a summer, but a thousand swallows, together with hot days, swimming, vacations, light clothing, and so forth, *do* make a summer. The first swallow or the first tulip is a sign of more to come. If more doesn't come we're disappointed. Dolly I could not possibly love me because when she revealed that she was a machine she implied that the long-term patterns of behavior that constitute love were absent from her repertory. I would also be disappointed if after hearing the first four notes of Beethoven's Fifth Symphony, I was told that that was all the orchestra was capable of playing—regardless of how well those four notes were played.

According to teleological behaviorism, love, like all other aspects of the human soul, is a complex pattern of behavior. Love is more complex than most patterns because it takes not just one person plus a social system but at least two people plus a social system. Of course, certain internal structures, certain neural organizations, are necessary for love to exist, just as an automobile has to be built in a certain way in order to perform in a certain way. But *Consumer Reports,* quite correctly, does not concern itself about how the automobile is built. *Consumer Reports* does not examine the

factory; they test the car. *Consumer Reports* is concerned with the automobile's actual *performance* (as well as its durability and serviceability). We
are consumers of love, not builders of lovers. Love is performance. The
idea that love and all mental life is performance, a behavioral pattern, is
the essence of teleological behaviorism. Teleological behaviorism does
not deny the validity of the physiological and the cognitive viewpoints.
The teleological view may be complementary to the others. Chapter 2 attempts to illustrate this complementarity with respect to one of them—
cognitivism—and to show that teleological behaviorism is no less "scientific" than either of them. To do this, we must first consider the roles of
efficient and final (teleological) causation in psychology and in science in
general.

Notes

1. Introspection as a method should be distinguished from the currently
active cognitive analysis of verbal reports; see, for instance, Ericsson and Simon
(1984).

2. British and Continental philosophy differed on many points, primarily on
what portion of mental life is due to innate categories, dispositions, or capacities
and what portion to associative processes. Wundt was influenced by Continental
philosophy which, following Kant, tended to be nativistic; Titchener's version
followed the direction of British philosophy—empiricist and associationistic. Neither, however, assigned much of a role to behavior and its consequences.

3. Aristotle's distinction between final and efficient causes (examined in detail
in Chapter 4) is at the core of this book's argument. Chapter 2 discusses final and
efficient causation in modern psychology. Roughly, an efficient cause is an answer
to the question, *How* does this or that event occur? An efficient-cause explanation
of a process typically takes the form of an underlying mechanism. A final-cause
explanation of a process is an answer to the question, *Why* does this or that event
occur? A final-cause explanation of a process typically takes the form of a goal or
purpose—the place of this particular process in its more abstract, more general,
more molar or wider context.

4. Jennings's position on mental terms is similar to that of teleological behaviorism. But Jennings, like Tolman after him, believed that the ascription of mental
life to nonhuman animals like the amoeba must be merely provisional; it had an as-
if quality. For Jennings true mental life, true consciousness, could only be known
subjectively. This belief is contrary to that of teleological behaviorism.

5. Tolman was somewhat influenced by E. B. Holt's (1915) operationistic
"neorealism." (See Smith, 1986, and Zuriff, 1984, for discussion of this movement.) Neorealists claimed that all mental life is "out there" in the world; they
conceived external objects and our relations to them as temporally extended patterns. Neorealistic operationism in some sense underlies teleological behaviorism.
But there are at least two reasons not to classify teleological behaviorism as merely
a form of neorealism: first, the neorealists, while explaining some mental states like
consciousness in terms of temporally extended patterns of behavior, identified
other mental states with external objects themselves. Pains and colors, for instance,
were treated by neorealists as characteristics of objects in the world (like chairs and
tables) rather than characteristics of behavior. Second, the neorealists were unclear

on scientific method. They rejected the materialistic "bead theory" of (efficient) causation but did not substitute any other.

6. Operationally, an internal state is nothing but a set of histories having equivalent effects. But neither Skinner nor his critics accepted such operationism. Both insisted that the concept of internal state implies actual internal mediation between environment and overt behavior, Skinner saying that such mediation is unnecessary in theory and his critics saying it is necessary.

7. Some early papers expressing this view are those of R. J. Herrnstein (1969, 1970), W. M. Baum (1973), and J. E. R. Staddon (1973). A recent special issue of the *Journal of the Experimental Analysis of Behavior* (Vol. 57, no. 3, May 1992) on "behavioral dynamics" contains several articles that fall more or less into this category.

8. According to the present categorization, cognitive neuroscience, which attempts to uncover particular physiological mechanisms underlying various cognitive states, is a form of physiologism.

2

Teleological Behaviorism
and Cognitive Psychology

Teleological behaviorism and cognitive psychology are two different approaches to a scientific understanding of the mind. The claim of cognitive psychology to be scientific—to be useful for predicting, controlling, and explaining behavior—rests on standard, well-understood scientific practice: observe behavior; form a theory, based on those observations, of an underlying mechanism; use the theory to predict future behavior; test the theory by further observation; revise the theory based on differences between observed and predicted behavior. The underlying causal conception is *efficient* causation—the familiar physical forces, each impinging on its successors. As I said previously, efficient causes answer *how*-type questions.

The claim of teleological behaviorism to be similarly useful rests on another less familiar form of causation: final causation. Final causes answer *why*-type questions. In order to understand the relation of cognitive psychology to teleological behaviorism, it is necessary to precisely distinguish final causes from efficient causes.

Aristotle's Concept of Cause

Aristotle's concept of cause was much wider than the modern one. A causal explanation of a process, for him, was an answer to a question about the process—whatever might follow the word "because" in a sentence (Hocutt, 1974; Randall, 1960, p. 124).

Aristotle refers to four types of causes: material, formal, efficient and final. Material and formal causes explain the nature of "substances," that is, static objects; efficient and final causes explain the dynamic behavior of objects, inanimate objects as well as organisms.[1] Aristotle's classification of causes may be most easily understood at this point by examples. The *material cause* of a substance, a circle for instance, is what it is made of. A circle might be made from a pencil line, a piece of cardboard, a piece of plastic (a bottle cap) or a mass of burning gas (the sun). The *formal cause* of a substance is an abstract conception of the class into which it fits. The equation $x^2 + y^2 = r^2$ would be a formal cause of a circle. All substantive circles have both matter and form, hence material and formal causes.

The *efficient cause* of an act, a girl watering a plant for instance, would be the set of internal nervous discharges giving rise to her muscular movements. *Final causes* of watering the plant might be the wider act of tending the plant or the still wider one of growing a garden. Chapter 4 will take up this four-part distinction in some detail. Now let us consider only the two dynamic causes, efficient and final.

In our modern way of thinking, causes precede their effects, and Aristotle's efficient causes do precede their effects. For instance, Aristotle says in *De Anima* (book II, chap. 12, 424a) that our sense organs are affected by forms of objects "the way in which a piece of wax [the organ] takes on the impress of a signet ring [the form of the object] without the iron or gold [the matter]." When discussing the effects of objects on *sense organs,* Aristotle does use efficient-cause explanations. But most of *De Anima* (On the soul) is devoted not to this subject but to the relation between objects and *whole organisms.* Such processes (to be discussed in detail in Chapter 5) are labeled with the familiar terms, "sensation," "perception," "imagination," and "thought" and are explained in terms of final rather than efficient causes.

Final causes vary in degree. The lowest degree of final cause is the action of a goal or an end on an instrumental act. When delivery of a food pellet depends on a rat's lever press, the rat may be said to press the lever not for the sake of pressing the lever but for the sake of eating the food pellet. J. L. Ackrill (1980) calls explicit goals of this kind "dominant." A dominant final cause (like the rat's eating a food pellet) may follow its effect (pressing the lever).

Higher degrees of final cause are called by Ackrill "inclusive" or "embracing." An inclusive final cause is to its effects as a wider act (like eating a meal) is to a narrower one (like eating an appetizer). Inclusive final causes are wider, more molar, more abstract than their narrower, more molecular, more particular effects. The narrower act is done not only for the sake of the wider but also for its own sake. Inclusive final causes fit into each other; eating an appetizer fits into eating a meal, which fits into a good diet, which fits into a healthy life, which in turn fits into a generally good life. The wider the category, the more embracing, the "more final" the cause. Ackrill (1980, p. 21) says: "Among ends all of which are final, one end can be more final that another: *A* [playing golf, say] is more final than *B* [putting] if though *B* is sought for its own sake (and hence is indeed a

final and not merely intermediate goal) [you might putt just to putt] it is
also sought for the sake of *A* [you might putt to play golf]"[2]

Analysis of inclusive final causes yields ends that consist of abstract
patterns of the movements that constitute them, ends that embrace those
patterns. Inclusive final causes are not simply efficient causes in reverse. An
effect of an efficient cause follows its cause but an effect of an inclusive
final cause does not *precede* its cause; it *fits into* its cause. True, a particular
movement must occur first in order for a pattern of movements to emerge
just as the movements of a symphony must be played before the symphony
can be said to have been played. In that sense and in that sense only an
inclusive final cause follows its effects.

Efficient causes may be thought of as answers to the question, *"How*
does this or that movement occur?" Analysis of efficient causes ultimately
yields "mechanisms" that may range from simple billiard-ball-like interactions to complex computer circuits to complex neurochemical processes
(Staddon, 1973). Because all efficient causes may in theory be traced back
to prior ones—what caused *them*—there is no ultimate efficient cause of
an act.

Final-cause analyses are attempts to answer the question, *"Why* does
this or that movement occur—for what reason?" Chapter 4 will argue that
for Aristotle the question "What reason?" is equivalent to, "Of what more
molar processes does this particular movement form a part?" Because
virtually any act may fit into a still-more-molar act (taking a bite of chick-
en, eating meat, eating a main course, eating a meal, sustaining one's
energy, surviving, reproducing) all final causes may in theory be expanded
in time to more-embracing ones.[3]

Answers to the question, "How?" regardless of their completeness do
not automatically answer the question, "Why?" According to Aristotle
(*Physics,* book II, chap. 7, 198a), it is the business of physicists to know *all*
the causes of their object of study. Since psychology, according to Aristotle, is a branch of physics, he would argue that psychologists ought to know
all the causes, final as well as efficient, of the behavior of organisms.

Aristotle's final-cause analysis of perception, memory, and creative
thought in *De Anima* and his discussion of freedom, responsibility, and all
of ethical behavior in *Nicomachean Ethics* follow one form: a particular act
is identified as a particular perception or thought, as free or unfree, as
good or bad, not on the basis of the particular *internal* acts (spiritual,
cognitive or physiological) that may efficiently cause it, but rather on the
basis of its environmental consequences or on the basis of the temporally
extended pattern of overt behavior into which the particular act fits—that
is, on the basis of its final cause.

As regards introspective knowledge of one's own mental states, one
Aristotelian interpreter says, "Aristotle has no reason to think that psychic
states—perceptions, beliefs, desires —must be transparently accessible to
the subject, and to him alone. Even if there are such states, this feature of
them is not the feature that makes them psychic states. Psychic states,

for human souls as for others, are those that are causally relevant to a teleological explanation of the movements of a living organism" (Irwin, 1980, p. 43).

Causation in Physics

For Aristotle, the motion of all physical objects is explicable in terms of final as well as efficient causes. A stone is perceived as naturally moving toward the center of the earth unless hindered. As long as the stone does indeed move toward the center of the earth, its movement may be explained in terms of its own ends. But if the stone should be thrown up in the air by a boy or a volcano, the movement of the stone would have to be explained according to ends other than those ascribable to it alone—those of the boy or the volcano. In that sense the stone was not "free." Talking about the ends of a stone's movement sounds odd to us because the success of Renaissance physics with its rigorous insistence on efficient causes has given us our modern conception of what does and does not constitute proper scientific language.

According to Richard Rorty (1982, p. 191):

> Galileo and his followers discovered, and subsequent centuries have amply confirmed, that you get much better predictions by thinking of things as masses of particles blindly bumping each other than by thinking of them as Aristotle thought of them animistically, teleologically, and anthropomorphically. They also discovered that you get a better handle on the universe by thinking of it as infinite and cold and comfortless than by thinking of it as finite, homey, planned, and relevant to human concerns. Finally, they discovered that if you view planets or missiles or corpuscles as point-masses, you can get nice, simple predictive laws by looking for nice simple mathematical ratios. These discoveries are the basis of modern technological civilization. We can hardly be too grateful for them. But they do not, pace Descartes and Kant, point out any epistemological moral. They do not tell us anything about the nature of science or rationality. In particular, they did not result from the use of, nor do they exemplify, something called "scientific method."

In this passage Rorty implies that in physics final-cause explanations have had less predictive power than efficient-cause explanations. Whatever their weakness in physics, final-cause explanations may have as much or more predictive power than efficient-cause explanations in psychology. Even in modern physics, final causes have a part to play.

According to Max Planck, the founder of quantum theory, the *"cause efficiens,* which operates from the present into the future and makes future situations appear as determined by earlier ones, is joined by the *cause finalis* for which, inversely, the future—namely a definite goal—serves as the premise from which there can be deduced the development of the processes which lead to this goal" (quoted in Yourgrau & Mandelstam, 1968, p. 165). If efficient causes have proved to be insufficient to explain physical phenomena at the most fundamental level, they may be no more sufficient in psychology. Therefore an attempt at final-cause explanation in psychology is worthwhile. (See Hocutt, 1974, and Silverstein, 1988, for corresponding arguments.)

Causation in Classical and Modern Psychology

Just as Renaissance physics replaced Aristotelian final causes with efficient causes, René Descartes (see Chapter 6) attempted to replace final causes of human and nonhuman animal behavior with efficient causes. According to Descartes, all nonhuman behavior is efficiently caused by external stimuli. In Descartes's model a stimulus acting through a sense organ is mechanically transmitted through the nerves to the brain whereupon animal spirits (supposed to be a material cause of life) are released to flow back through the nerves to expand the muscles and cause movement. The difference between humans and other animals is that in addition to this completely automatic mechanism, which causes involuntary behavior, humans have another way to direct animal spirits to various nerves—their *will* which, acting through the pineal gland in the brain, can alter the direction of the flow of animal spirits. A person's behavior whether voluntary or involuntary is always, according to Descartes, efficiently caused. Voluntary behavior is efficiently caused by the will, while involuntary behavior is efficiently caused directly by external stimulation. Thus Descartes brought Renaissance psychology into line with Renaissance physics.

Contrary to Aristotle's conception of the nonprivacy of mental states (see previous quote from Irwin, 1980), Descartes held that the essential fact about mental states is their privacy. "I think, therefore I am," is Descartes's pronouncement about what was "clear and distinct" in the privacy of his own mind. The existence of anything else—his own body, its behavior and the bodies and behavior of all other things—rested, for Descartes, on this internal and private clarity.

Since Descartes, psychology has, in various ways, attempted to demystify the action of the will. One method has been to incorporate all behavior, voluntary as well as involuntary, into more or less complicated reflex systems. I. M. Sechenov's (1863/1965) *Reflexes of the Brain* and Pavlov's (1927) *Conditioned Reflexes* represent such systems. According to Sechenov (from Herrnstein & Boring, 1965, p. 321), *"the initial* [efficient] *cause of all behavior always lies not in thought, but in external sensory stimulation, without which no thought is possible"* (italics in original).

The American behaviorists, Watson (1913), Hull (1952), and Tolman (1949) differed strongly on how to describe the internal efficient causes of behavior but their ultimate object, no less than that of Sechenov and Pavlov, was to explain behavior in terms of its internal efficient causes, to get behind behavior itself to discover its underlying mechanisms. Disputes between these behaviorists and between behaviorists and cognitive psychologists, as well as between various schools of cognitive psychology, have centered on the question, What are the internal efficient causes of behavior? At one extreme, Sechenov, Pavlov, and their modern descendants (for example, Rescorla, 1988) attempt to trace the "reflexes of the brain." The connectionists, E. L. Thorndike (1911) and Hull (1952), spoke more abstractly in terms of internal stimulus-response connections or associations perhaps eventually reducible to physiological reflexes.

More molar behaviorists like Tolman (1949) abandoned the possibility that mental constructs could be reduced to reflexes but still retained the goal of describing internal constructs (if not mechanisms). In Tolman's hands S-R psychology became S-O-R psychology, where the O was conceived as a set of efficient causes *mediating* between environment and behavior. The cognitive psychologist John R. Anderson (1991, p. 513) says: "I have always felt that something was lost when the cognitive revolution abandoned behaviorism." But he goes on, "In doing this, however, I do not want to lose the cognitive insight that there is a mind between the environment and behavior."[4]

For some modern philosophers of psychology a given mental state may be internally represented by the action of a given computer mechanism consisting of individual components, none of which may actually represent the mental state. Daniel Dennett (1978) calls such mechanisms "intentional systems." As a very simple example of such a system Dennett (1978, pp. 71–89) cites Thorndike's law-of-effect. Figure 2.1a shows a simple cognitive model of the law of effect. In the operation of the law of effect, in Thorndike's original conception and Hull's (1952) elaboration, reinforcement strengthens a modifiable S-R connection causing behavior that appears purposive, despite the fact that the organism's purpose itself has no coherent internal representation. Another example of an intentional system would be a chess-playing computer that, say, had a tendency to bring out its queen too soon even though no such tendency was explicitly programmed in the machine. Mental states, according to this sort of cognitive psychology, may be *emergent* qualities of behavior. Research on "neural networks" (Grossberg, 1982) is perhaps the clearest example of how complex cognitive and behavioral processes may emerge from the concatenation of much simpler efficient causes.

Another sort of modern philosophy of psychology (Fodor, 1981) insists that all mental states are internally represented as such and interact with each other in the mind to cause behavior. According to Fodor (1981, p. 5), "Mental causes typically have their overt effects *in virtue of their interactions with one another*" (italics in original). *Desire* for food, for instance, interacts with *knowledge* of how to get it. Psychology would consist of the analysis of such interactions. Philosophers of this latter school frequently (and not without a degree of justification) accuse those of the former of being behaviorists. The issue within classical behaviorism as well as between classical behaviorism and modern cognitive psychology (of either school) is not whether psychology consists of the analysis of an efficient-cause mediating process between environment and overt behavior; all follow Descartes on the necessity of an efficient-cause analysis. The issue is the role of mental terms in such an analysis.

In general, aside from Skinner, behaviorists and cognitivists alike have viewed psychological theory as about efficient causes *mediating* between environment and behavior. To Skinner we owe the renaissance of the Aristotelian focus on the behavior of whole organisms in their environments.

Causation in Skinnerian Psychology

In the more than half-century between his first published paper and his death in 1990, Skinner naturally varied his views considerably. It is not my object here to trace that development. One thread that runs throughout his work, however, is his ultimate concern with behavior of the organism as a whole. The concept of the *operant* as a class of individual movements of whole organisms having a common effect, the causal power attributed to *contingencies of reinforcement* of operants, and the weight assigned to an organism's and a species' *history of reinforcement*—these are strains of Aristotelian thought in Skinner's psychology.

With regard to the Thorndike-Hull model of Figure 2.1a, Skinner objected to what reinforcers were supposed to do: strengthen S-R connections inside the animal. The neurophysiologist Sherrington (1906) had emphatically and demonstrably reiterated Dewey's (1896) argument that reflexes do not exist as such inside an animal. A stimulus has, according to Sherrington, no coherent internal representation (still less does a "situation"). The path of stimuli through the nervous system, Sherrington claimed, is diffuse. However, somewhere near the point at which a response is triggered, the diffusion is brought into focus in what Sherrington called "the final common path." Skinner (1938) analyzed available evidence to show that not even the final common path of a reflex had a coherent internal representation. A reflex response, according to Skinner, comes to focus only in *overt* behavior; the final common path of a reflex is the response itself (see Fig. 2.1b). A reflex, according to Skinner, is therefore a completely overt phenomenon, a correlation between an external stimulus and an overt response. If, as Skinner claimed, no coherent internal representations of stimulus-response connections exist, there can be no reinforcement, no strengthening, of such connections. According to Skinner, reinforcement strengthens not an internal S-R connection but (through a complex neural network) the overt response itself.

As noted previously, an individual reinforcer may be conceived as a final cause (a dominant final cause) of the instrumental acts upon which it is contingent. However, while Skinner was never deeply concerned with the logic or philosophy of the concept of causation, there are signs that he was not comfortable with teleological explanation as such. Skinner (1938, p. 69) held that an individual instance of reinforcement served as the

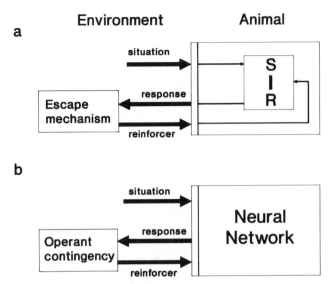

Figure 2.1   (a) The Thorndike-Hull model of reinforcement of S-R connections as an explanation of how a cat learns to escape from a "puzzle box." (b) Skinner's conception of the direct reinforcement of responses.

efficient cause of subsequent response-rate increases. His concept of "superstition" rests on adventitious contiguities as (efficient) causes of future behavior. Even Skinner's most molar construct, "history of reinforcement," is construed as existing *wholly* in the past. Skinner saw a person's reinforcement history as the efficient cause of her subsequent behavior rather than the (inclusive) final cause of the behavior constituting the history.

Aristotle, as part of his discussion of nutrition and reproduction (*De Anima*, book II, chap. 4, 414b), makes a distinction between the soul as an efficient cause and the soul as a final cause. The soul may be the efficient cause of behavior only in the sense that parents, by their reproductive activity, pass their behavioral characteristics to their offspring. That is, the reproductive (the most basic) part of the parents' souls efficiently causes their offspring's souls. But within a person's lifetime his or her soul is the *final*, not the efficient, cause of behavior. An individual's "history of reinforcement" (Skinner's concept closest to Aristotle's "soul") thus would have been viewed differently by Aristotle and Skinner. For Skinner it was an efficient cause of behavior. Aristotle would have viewed it not as an efficient cause, preceding any particular act, but as a final cause embracing the act.

Skinner's conception of reinforcement has seemed to some modern commentators (Ringen, 1985; Staddon, 1973) to involve efficient causation—but with temporal gaps. An individual instance of reinforcement may immediately increase response rate but then what stands between one response and another? Certainly a history of reinforcement, considered as an efficient cause of temporally remote behavior, must act through some (presumably internal) mechanism. In referring to this possibility, J. E. R. Staddon (1973) says: "I suggest that the apparent simplicity of the relation between operant behavior and its consequences implies not an absence of mechanism, not that operant behavior is 'undifferentiated material,' in Skinner's phrase, but rather the existence of more and richer mechanisms than have hitherto been seriously contemplated [p. 55]. . . . Explanations in terms of 'purpose' or 'final cause' are always incomplete" (p. 58). True. But Staddon does not add that explanations in terms of efficient cause ("mechanisms" in Staddon's terms) are also incomplete. Good automobile mechanics are not necessarily good drivers; they may not be able to predict and control the movements of the car as a whole.

The price Skinner paid for the rejection of teleology was a simultaneous rejection of *mental* terms and acceptance of private *inner* causes (Zuriff, 1979). Willard Day (1969) referred as follows to an exchange between Skinner and the philosopher Michael Scriven: "Skinner is objecting . . . not to things that are private but to things that are mental." Here is Skinner's (1953, p. 279) not atypical interpretation of an idea: "If the individual himself reports 'I have had the idea for some time but have only just recently acted upon it,' he is describing a covert response which preceded the overt." Chapter 5 will attempt to show that for Aristotle, an idea is not a covert response at all, but a pattern of wholly overt responses including the individual's verbal report as one particular part of the pattern.

In relegating mental concepts such as ideas to the interior of the organism (the same place where the organism's physiology exists), Skinner is in line with modern cognitive psychology. Both Skinner, in the above passage, and much of modern cognitive psychology place mental and physiological events together ("cognitive neurophysiology") inside the organism. The difference is that for Skinner such internal events are unimportant simply because they *are* internal events, hence not controlling variables (1953, p. 35):

> The objection to inner states is not that they do not exist, but that they are not relevant in a functional analysis. We cannot account for the behavior of any system while staying wholly inside it; eventually we must turn to forces operating upon the organism from without. Unless there is a weak spot in our causal chain so that the second link is not lawfully determined by the first, or the third by the second, then the first and third links must be lawfully related. If we must always go back beyond the second link for prediction and control, we may avoid many tiresome and exhausting digressions by examining the third link as a function of the first. Valid information about the second link may throw light upon this relationship but can in no way alter it.

For modern cognitive psychologists, on the other hand, internal events are critically important; their study is the essence of psychology itself. Dennett (1978) distinguishes between "sub-personal" level theories (theories of within-person mechanisms) and "personal" level theories (theories of the behavior of the person as a whole). According to Dennett (1978, p. 154, footnote), "the personal level 'theory' of persons is not a psychological theory."[5]

Skinner (1938) claimed that there are only two basic kinds of behavior: *respondents,* classes of behavior (like a person's pupillary dilation or a dog's salivation) elicited by immediate antecedent stimulation, and *operants,* classes of behavior correlated with immediate environmental consequences. Examples of operants are rats' lever presses, pigeons' key pecks, and all animal behavior normally considered voluntary. Aspects of the environment crucial for respondent and operant dynamics are antecedent *stimuli* of respondents, *reinforcers* of operants—consequences that generally increase an operant's rate of emission (operant conditioning being the study of the relationship of operants and reinforcers)—and *discriminative stimuli,* in the presence of which a given operant-reinforcer relation obtains. An example of a discriminative stimulus is the open / closed sign on the door of a shop, signaling a given relation between door-pushing (the operant) and door-opening (the reinforcer).

Nowhere here are mental terms. Sometimes Skinner (1953) offered "interpretations" in which the use of mentalistic vocabulary in everyday speech is explained in terms of operants, reinforcers, and discriminative stimuli. Self-control, for instance, is held to be nothing but operant avoidance of certain discriminative stimuli ("get thee behind me, Satan"), like crossing the street to avoid the enticing smell of a bakery.

Skinner's nonmentalistic terminology has served very well in the analysis of discrete operants like pigeons' key pecks, rats' lever presses, and humans' button pushes (Honig & Staddon, 1977). Furthermore, the patterns of behavior discovered in one situation with one species often appear in other situations with other species. These patterns change in systematic ways with motivational variables like reinforcer deprivation and drug dosage. Skinnerian techniques have been successful in areas of behavior therapy, ranging from treatment of severe psychoses to weight control; their great advantage in these applications is their resolute focus on *consequences,* that is, contingencies of reinforcement. For example, many people suffer from *agoraphobia;* they are house-bound, have panic attacks in public places, and consequently refuse to leave home. In searching for causes and treatment of such behavior the Skinnerian behavior therapist considers not just its antecedents but also its consequences: avoidance of work, avoidance of social responsibility, avoidance of sexual temptation, attention from relatives and friends, etc. Focusing on the actual consequences of dysfunctional behavior has led in many cases to the development of successful treatment by substitution of less dysfunctional behavior to achieve equivalent ends. Skinnerian techniques have been successfully applied also in business management and in areas normally considered mentalistic, like the teaching of reading and mathematics to children and in college-level courses as diverse as anatomy and foreign languages.

However, despite this success, it has not been possible either in the operant laboratory or in the many areas of psychological application to divide all behavior neatly into specific respondents and specific operants. A respondent must be correlated with a particular antecedent stimulus and an operant with a particular consequent reinforcer. What, for instance, reinforces the act of refusing an offered cigarette by a smoker trying to quit? Having to deal with and talk about such obviously important acts, behavior therapists have taken two roads, neither satisfactory.

Some, like L. E. Homme (1965), have developed an operant psychology of the hidden organism, speaking of inner (covert) respondents, inner operants ("coverants"), inner reinforcers, and inner discriminative stimuli. According to these psychologists the person who refuses an offered cigarette can just reinforce the act himself (pat himself on the back, so to speak). This conception has both logical and empirical problems. Logically, if a person can reinforce his own actions why should he ever withhold reinforcement of *any* action? What reinforces the giving and withholding of self-reinforcement (Catania, 1975)? Empirically, there is just no evidence that self-reinforcement works and some evidence that it doesn't work (Castro & Rachlin, 1980).

The other road taken by behavior therapists has led to cognitive behavior therapy (Mahoney, 1974). Cognitive behavior therapists retain Skinnerian techniques for acts that are clearly reinforced. But, where environmental reinforcers are not obvious or immediate, cognitive behavior therapy abandons behaviorism entirely and refers to mental states as inner causes. Thus the person who refuses the cigarette may do so because he *believes* it is better for his health and because he *wants* to be healthy. A therapist might then try to strengthen the person's belief and desire by logical argument or by asking him to repeat a statement of his belief over and over or by reinforcing the statement of the belief. Even this last procedure is cognitive, not behavioral, because it rests on the assumption that the statement is merely evidence of an internal state and that the reinforcement acts not only on the external statement but also on the inner belief. After all, it is the refusal of cigarettes (what the belief is said to cause) rather than the verbal statement that the therapist is ultimately trying to strengthen. In principle, there is nothing wrong with cognitive behavior therapy. If people do have beliefs as inner states and if beliefs can cause specific actions, then changing the belief will change the action.

From a viewpoint that sees mental terms as descriptions of inclusive final causes, however, the cognitive behavior therapists are making what Gilbert Ryle (1949) calls a "category mistake." If a belief is nothing but a pattern of actions, then a statement of a belief is merely one of those actions. Altering the statement would affect the belief only to the extent that it affects one part of the pattern, not the central source of all parts. Beethoven's Fifth Symphony, for instance, is a pattern of notes with four very familiar notes at the beginning. When you hear those notes from an orchestra you can be pretty sure that you are about to hear the rest. To alter those four notes is to alter Beethoven's Fifth Symphony seriously (perhaps to the extent that you would want to call it something else) but not necessarily to have any effect on the other notes.

But then again, there may indeed be a central state, more or less coherently represented in the nervous system, more or less innate, that controls all the behavior that an outside observer would call evidence of a person's belief. Altering a belief (say, substituting the score of Schubert's Ninth for Beethoven's Fifth) would then alter *all* of its behavioral effects. Cognitive behavior therapists are thus trying to get at the central antecedents, the *efficient cause,* the core, the nub, the nut, the origin, the control room of belief behavior, the very essence of a person's belief. The problem is that they have abandoned the aspect of Skinner's program that made it so successful: its concentration on consequences rather than antecedents. A therapist who focuses on the central efficient causes (the *how*) of a person's belief tends to lose sight of the reinforcers—the person's relations with his family, friends, with his environment in general. In other words, the *why* of the belief.

The failure of behavioral psychology to deal with mentalistic concepts such as beliefs (which philosophers call "propositional attitudes" or "intentional states") and pains (which they call "phenomenal states" or "raw feels") has by and large left treatment of strictly mental dysfunction to non-behavioral psychology. A comprehensive behavioral analysis of such concepts does not currently exist but such analyses are possible and may be highly useful. (See Chapter 7 for a discussion of pain as behavior.)

Teleological Behaviorism

E. R. Guthrie (1935) distinguished between *acts* and *movements*. A movement, according to Guthrie, is a particular set of muscular contractions resulting in a particular locus of bodily displacement. (Guthrie's definition of movement is thus much narrower than Aristotle's). An act, according to Guthrie, is a coordinated pattern of movements leading to some definable result. Waving goodbye, for example, would be an act. Ten different instances of a person's waving goodbye might involve ten wholly different sets of movements but constitute a single act, repeated ten times. Guthrie claimed that the apparent learning of acts was essentially an accident of the learning of particular movements. Skinner's (1938) concept of operant conditioning reversed Guthrie's claim. An operant (an act as Guthrie defined it) may be directly reinforced, according to Skinner, without regard to particular movements.

Teleological behaviorism expands Skinner's original concept of reinforcement from a single event dependent on a single operant—for example, a single food-pellet delivery immediately following a single lever press—to a pattern of environmental events perhaps only vaguely contingent on an overlapping pattern of operants. In R. J. Herrnstein's (1969) conception of avoidance, for instance, the effective contingency is a negative correlation imposed between the rate of aversive events and the rate of operant emission, not any individual consequence (hypothetical or real) contingent on any particular operant. Even so abstract a conception as *belief* is such a pattern of overt observable acts and consequences. Take away the observable acts and the belief goes away as well; it is just as if you were to take away the lights and shadows on the movie screen. Without lights and shadows, the action, the characters, their motives, their beliefs, all go away as well.

A given act may be truly understood only some time, perhaps a considerable time, after it occurs because the context (the final cause) of an act extends into the future as well as into the past. An individual lever press is uncaused in exactly the same sense that an individual event has no probability. A probability (conceived as a relative frequency) may be known as accurately as you want it to be known if you are willing to wait. The same goes for the causes of an individual lever press and, arguably, the same goes for a person's belief (including one's own belief). If a behavioral analysis ultimately fails, the failure will be due to the complexity of the task. It will be like the failure to predict and control the weather precisely rather than any intrinsic inaccessibility or opacity of its subject.

Probability

The difference between cognitive (efficient-cause) psychology and teleological (final-cause) behaviorism is perhaps most clearly illustrated in terms of their separate conceptions of probability. Figure 2.2, complicated as it is, is a simplified picture of how cognitive and behavioral models deal with identical concepts. (For the sake of brevity, we drop the tag "teleological" and just say "behaviorism," "behavioral," and so forth.) The double vertical line represents the boundary between a person and the world. Note that five horizontal lines, three solid, two dashed, cross the boundary. The solid lines are the three critical behavioral variables. The top solid line heading into the person represents *data* or information of some kind, including such stimuli as train whistles, red and green traffic lights, or the instructions a psychologist might give to an experimental subject. These informative stimuli may function in two ways: they may signal significant *outcomes* (like trains signaled by whistles or like food signaled by tones in Pavlov's experiments). In such cases these signaling events are called *conditional stimuli* or CSs. The relation of CSs to significant environmental events (*unconditional stimuli* or USs) is independent of an animal's behavior. In Skinnerian terms this relation is a *respondent contingency*. The whole process, including its effects on behavior, is called *respondent conditioning*.

Alternatively the information could signal not a significant event as such but the relation between *behavior* and *outcome*, a relation called, again

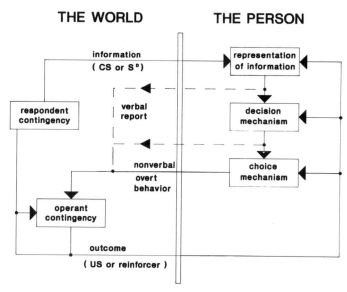

THE WORLD   **THE PERSON**

Figure 2.2 A behavioral model (to the left) and a cognitive model (to the right) of probabilistic choice.

using Skinner's terms, an *operant contingency*. Informative stimuli that signal the presence of an operant contingency are called *discriminative stimuli* (or S^Ds). The red and green traffic lights that signal the relation between crossing the street and likelihood of having an accident (or getting a ticket) are discriminative stimuli. This whole process, including its effects on behavior, is called *operant conditioning*. (There is much debate among behaviorists, which we shall completely ignore here, whether respondent conditioning is really a form of operant conditioning, or vice versa.)

The two dashed arrows crossing the world-person boundary represent verbal reports (of representations and decisions in the illustration). Verbal reports may be complex and temporally extended (as are many nonverbal actions such as playing a saxophone), in which case (like other complex overt actions) they have a complex syntactical structure. Alternatively they may be simple and discrete (grunting or saying "ouch" or "yes" or "no"), in which case they may have a simple structure. In either case verbal reports like other overt behavior may be elicited by a stimulus (a US) or entail significant consequences, thus taking part like other overt behavior in respondent or operant conditioning.

The behaviorist studies the relation between these boundary-crossing events *from the left* in Figure 2.2. Behavioral inferences and models are inferences and models about respondent and operant contingencies that may not be present at the moment but serve as the *context* for current actions (Staddon, 1973). A grocer, for instance, may give a poverty-stricken customer a free loaf of bread because the grocer is at heart a sympathetic and generous person or as part of a promotional campaign.

The grocer's differing motives form differing potential contexts for (and causes of) his act. From a behavioral point of view these contexts differ by virtue of differing respondent and operant contingencies operating over periods of time much wider than the present moment. The question of which motive the grocer *really* has can be answered, from a behavioral viewpoint, *only* by reference to such overt contextual events—the existence or nonexistence of a promotional campaign and the grocer's behavior when no such promotional campaign is being waged.

One important feature of the behavioral viewpoint bears emphasis: the question may be settled by reference to future as well as to past events; the temporal context of a brief event extends into the future as well as the past. To decide what the grocer's motives really were you could do an experiment post hoc: say, sending poor customers into the store on different occasions. (In technical terms, you would be trying to determine a utility function for the grocer in order to predict his behavior in other circumstances.) If the grocer should die immediately after his generous act (and historical records are lost or nonexistent), there is *no way, even in principle,* to determine its behavioral context (the grocer's motives). Supposing (probably contrary to fact) that Beethoven's Fifth Symphony and Schubert's Ninth Symphony had three identical successive notes, it would be like trying to determine from those three notes alone which symphony you were listening to.

To repeat again, from the behavioral viewpoint, a mental act such as a motive has meaning only over an extended span of time. To sample only a piece of that time is to make a guess as to the motive. As wider and wider samples are taken the grocer's motive becomes more and more knowable but it will never be 100% knowable; in principle, its context is infinite. This is true even with respect to the grocer's knowledge of himself, because from the behavioral viewpoint (from the left in the diagram) the grocer is in a privileged position only by virtue of the quantity rather than the quality of his information: he has a larger sample than anyone else of his own behavior but by no means a better sample. Indeed, his information as to his own motives may be much worse than that of an outsider because he can only observe his own behavior (as a whole person) by reflection (through its various outcomes), whereas outsiders may view his behavior directly. Thus the behaviorist turns on its head the common mentalist notion that people have privileged access to their own mental states. In this respect, teleological behaviorism is like Freudian psychology (although of course different in many other respects).

Cognitive psychology takes the same sorts of data as does behavioristic psychology—the five arrows crossing the world-person boundary—and infers a context of those events not in past and future contingencies but in the present, inside the behaving organism. Cognitive psychology looks at the diagram of Figure 2.2 from the right. The cognitive context of the grocer's generous act therefore consists of events inside the grocer (or states of internal mechanisms) contemporaneous with (or immediately preceding) that act—its efficient causes.

The right section of Figure 2.2 illustrates in highly schematic terms how a cognitive model (Kahneman & Tversky's, 1979, prospect theory) handles probabilistic information. In a typical experiment each of a large group of subjects may be asked a series of hypothetical questions of the following form: "Which would you prefer, a 50% chance of winning $10,000 or $5,000 for sure?" (By far the greater number of subjects say they prefer the sure thing.) The first stage of the model consists of "editing" the problem and transforming its four elements (probabilities of 1.0 and .5, amounts of $5,000 and $10,000) into internal representations. The internal representations may be reported directly (upper dashed arrow), but their creation is only the first step of the decision process. They are then combined in some way specified by the theory to arrive at a decision and the decision is reported (lower dashed arrow). Investigators have found verbal reports of representations of probability to be generally unreliable predictors of decisions (the effective internal representations of probability are termed "decision weights" in prospect theory). But verbal reports of decisions are quite reliable predictors of actual choices when (occasionally) subjects are asked to choose among actual rather than hypothetical probabilistic outcomes. (Of course, money amounts are much smaller in the actual choice experiments.) The ultimate object of a cognitive theory is to predict decisions and choices in a wide variety of real-life and laboratory situations. And, arguably, prospect theory does a good job.[7]

The important point for us is that cognitive theory conceives probability as an internal state. According to prospect theory (and most cognitive theories), the experimenter who says, "The probability of $10,000 is .5," is merely activating an internal representation. The true probability is that representation. This reflects a *subjectivist* view of probability, as a degree of confidence or degree of certainty about coming events. That confidence or certainty then may determine subsequent choices (Lucas, 1970). When a weatherperson says, "The chance of rain is 90%," the confidence in rain thus aroused may cause the listener to take his or her umbrella to work. The weatherperson's words are efficient causes of the confidence, which is (part of) the efficient cause of the decision to take an umbrella, which in turn is (part of) the efficient cause of actually taking the umbrella. The critical mechanism may involve various feedback loops (as Figure 2.2 shows) and may well be much more complex than this (as even a knee-jerk reflex is much more complicated than a simple series of S-R connections), but it is nonetheless a mechanism, one that involves a series of efficiently caused acts.

For a behaviorist, the experimenter's assertion, "The probability of $10,000 is .5," does not elicit an internal representation; it already *is* a representation of a class of wholly external probabilistic events. The function served by the verbal statement is to place the present narrow situation (the psychology experiment) in the context of those events. In other words, the probability statement is a discriminative stimulus for a particular history of reinforcement. Thus, to say to a person that the probability of X is .5 is to say something like, "Behave in this experiment as if a coin were being flipped and you were rewarded with X if the coin came up heads." Unless the person has had that sort of reinforcement history, the experimenter's statement is meaningless. The true probabilities, from a behavioral viewpoint, are the relative frequencies of events themselves. This is an *objectivist* view of probability. The weatherperson's statement, "The probability of rain is 90%," is itself, for the behaviorist, a discriminative stimulus for taking an umbrella to work and is established, in the same way that any discriminative stimulus is established, by reliably having signaled a given set of operant contingencies. On previous occasions in this general context, when this sort of weather was predicted, acts classified as heeding the prediction (taking an umbrella to work, for instance) were reinforced and acts classified as failure to heed the prediction were punished. In other contexts—unreliable weatherpersons, for instance—the same specific prediction would have a different meaning.

For the behaviorist, the essential meaning of a probability statement is what it represents externally (its discriminative function), while for the cognitivist the essential meaning of a probability statement is how it is represented internally. This is a difference between behavioral and cognitive *semantics*. Note, however, that neither the behavioral nor the cognitive approach deals with or accounts for *syntactic* structures. It is no more clear why several phonetically different statements should arouse the same internal representation than it is why several phonetically different statements should serve as discriminative stimuli for the same overt choice behavior.

One reason for discussing behavioral and cognitive approaches to probability here is that experimental work with both approaches seems to have converged on corresponding models (Rachlin, Logue, Gibbon, & Frankel, 1986; Rachlin, 1989). I have no space to trace out all of the various correspondences but to get their flavor the reader might imagine how one would present a person directly (that is, nonverbally) with a probabilistic event. It is impossible to present a probability in an *instant*. A coin has to be flipped many times for the probability of .5 of heads to emerge as a property of coin flips. Thus a probability itself must be presented as a relative frequency of actual events. As it happens, the choice behavior of humans and nonhumans in the face of various relative frequencies of actual events corresponds to human verbal reports of decisions in the face of verbally presented probabilities; the form of the functions relating probability to behavior is the same for humans and nonhumans, accounting for the same sorts of apparent irrationalities, while the parameters of those functions differ widely between species.

Teleological Behaviorism and Modern Philosophy of Psychology

At this point it may fairly be asked why, if Aristotle's conception of final causes is consistent with behaviorism, are modern philosophers, particularly those who consider themselves Aristotelians, so dead set against behaviorism? The answer is not that these philosophers misunderstand

Aristotle; they do, however, misunderstand behaviorism, especially its molar character, and they completely ignore teleological behaviorism. However, one point of agreement between modern Aristotelian philosophy and behaviorism is a mutual rejection of introspection and the private nature of mental events. Alan Donagan (1987, p. 54) says: "A false and deeply confused doctrine that was philosophically fashionable is still encountered: namely, that taking propositional attitudes [like beliefs] and persisting in them are items in one's flow of private consciousness, which are named by private ostensive definition, and which have complex causal relations with one's bodily states."

Donagan seems to be arguing against introspection as a method in psychology. Aristotelian philosophers in general do indeed deny the privileged status of introspection.[8] But Donagan goes on: "[I]t does not follow that description of actions in terms of their doers' propositional attitudes can be analyzed without residue in terms of patterns of their surface behavior" (54). Let us discuss that residue. Donagan does not say that if you removed all of the surface behavior you could still have the residue left. Rather, he implies, when we look at the surface behavior, we perceive it and describe it in such a way that our perceptions and descriptions cannot be analyzed into strictly behavioral terms—just as you could not analyze the personality of the characters in a movie into lights and shadows. Then, one must ask, how *do* we perceive beliefs and other propositional attitudes? The answer, according to Donagan and most contemporary philosophers, Aristotelians and others, is that we perceive them as arising from inside of the person, ourselves or someone else, who has them.

Fred Dretske (1988, p. 2) says, "behavior is endogenously produced movement, movement that has its causal origin *within* the system whose parts are moving" (italics in original). Thus, according to Dretske, acts to which no internal cause is attributable are not even behavior let alone voluntary behavior. For Dretske, overt behavior as we perceive it comes with a sort of (efficient) causal tail. A rat's lever press, if it is perceived as behavior, is perceived as being caused by something inside the rat. The lever press itself *plus* the cause constitute the rat's act. Correspondingly, when we perceive a person's actions as arising from that person's belief, we also perceive behavior with a tail extending inside the person. But this time the tail wags the person, so to speak. The part of our perceptions that reaches inside the person contains not just a single internal efficient cause, as in the case of a rat's lever press, but an entire representational system. The person's cognitive state consists of the overt act and the internal representational system together. (Dretske's analysis of behavior is thus essentially cognitive.) The representational system (the *reason* for the act), like the causal tail on the rat's lever press, is the "residue" inevitably left over from any conceivable analysis of surface behavior alone. This is how modern philosophy justifies modern cognitive psychology: internal states as either efficient causes of or less-determinate "reasons for" external behavior.[9]

But the present discussion of Aristotle's philosophy suggests that the

residue after the most penetrating finite behavioral analysis is just more
behavior—like the residue of uncertainty about the probability of a coin
being unbiased after x number of tosses. There is inevitably a residue in
this case, but it exists because an ideal analysis would take an infinite
amount of time, not because the residue exists in another place.

There is no question that sometimes, in our society, in our linguistic
environment, we attach inner causes to external behavior. Such attach-
ments may harmlessly be called a branch of "folk psychology," which then
may be said to have developed into modern cognitive science (Stich,
1983). But as J. R. Kantor (1963) argued, there is another branch of folk
psychology, a sort of folk behaviorism, that is only satisfied when an act is
explained in terms of its consequences. Our folk psychology tells us to be
more satisfied with an explanation of a murder, for instance, in terms of
consequences like money or sex than in terms of inner compulsions. When
a man kills his wife and then cashes in her insurance policy and runs off
with his mistress, we do not forgive him when he says he could not resist
his inner compulsions. It is only when there is no clear and direct environmental consequence that we are driven to take inner causes seriously. But
since in such cases we try to discover a plausible internal representational
system, as cognitive psychology is set up to do, why is it not just as valid to
try to discover a plausible environmental consequence, even if unclear and
indirect? This, it will be shown, is exactly what Aristotle did do in corresponding circumstances when he looked for final causes as behavioral
explanations.

Kantor (1963) claimed that a behavioristic folk psychology is the true
one; indeed, the only one on which a scientific psychology could be built.
Why then, it may fairly be asked, would we *ever* in our everyday mutual
interactions or our technical discourse talk as though mental events were
internal efficient causes of our behavior? Charles Taylor (1964) blames it
on "atomism," which he defines (p. 11) as "the notion . . . that the ultimate evidence for any laws we frame about the world is in the form of
discrete units of information, each of which could be as it is even if all
others were different." On this basis, Taylor criticizes both modern cognitive psychology ("centrism") and classical behavioral psychology ("peripheralism"). Taylor (p. 12) says: "Teleological explanation is, as has often
been remarked, connected with some form of holism, or anti-atomistic
doctrine."

However, the sort of holism Taylor advocates differs from the behavioristic holism advocated here. Taylor's (and recently Dretske's) holistic
approach embraces an organism's current *internal* as well as its current *overt*
behavior (peripheralism *plus* centrism). The teleological behaviorist's holistic approach, on the other hand, embraces an organism's *past* and *future*
overt behavior as well as its *current* overt behavior. The former is a holism
of space; the latter, a holism of time. The present analysis therefore agrees
with modern Aristotelian philosophers that mental terms (propositional
attitudes, dispositional states) form at least part of our fundamental "data

language." The issue is what to do with those data. Should they be analyzed into current overt behavior plus internal causes and reasons, as some part of our folk psychology seems to dictate? Or should they be analyzed wholly on the level of overt behavior and its consequences over time, as some other part of our folk psychology seems to dictate? This chapter has merely reiterated in modern terms Aristotle's answer: both avenues need to be pursued. We should know *all* of the causes of our object of study; success in either, far from blocking the other, will illuminate its path. The next several chapters will again take up the question of how the mind may be understood—but in its original form as addressed by the ancient Greek philosophers, Plato and Aristotle themselves.[10]

Notes

1. Unless otherwise noted, references to Aristotle and quotations from Aristotle's work refer to Richard McKeon's (1941) edition. A later edition (Barnes, 1984) is available but that edition, though more complete, is less readable and less consistent in terminology than the earlier one.

2. Economic utility functions assign values to all conceivable "packages" of activities, including instrumental and consummatory activities. When the value of a consummatory activity (like eating by a hungry rat) is far higher than that of an instrumental activity (like lever pressing by a rat not deprived of exercise), then as in Ackrill's (1980) teleological analysis, the reinforcer may be said to "dominate" the instrumental act. But the case of teleological dominance is just one extreme on a continuum of relative value of a pair of activities constrained in a certain ratio (Timberlake & Allison, 1974; Premack, 1965). Lever pressing is always to some extent performed for its own sake as well as for the sake of eating. Thus, according to modern reinforcement theory, including the economic theories discussed here, "dominant" final causes are just lower forms of "inclusive" final causes. Chapters 4 and 5 will argue that Aristotle's own concept of final cause is also fundamentally inclusive.

3. Modern religious thought tends to identify the deity with an ultimate efficient cause, whereas Aristotle identified the deity with an ultimate final cause. St. Thomas Aquinas's resolution of Aristotelianism with Christianity in the late Middle Ages may be conceived to include a view of God as both ultimate efficient and ultimate final cause of all movement.

4. If the word *between* were held to imply a relationship (like an agreement *between* two people) rather than a third object (like the filling *between* two slices of bread in a sandwich), Anderson's statement could be interpreted as an affirmation of teleological behaviorism.

5. Dennett does not say exactly what the scientific status of a personal level "theory" is. By putting the term in quotes, Dennett perhaps implies that not only is a personal level theory not a psychological theory but it is not a scientific theory at all; it is somehow worth less than a subpersonal theory. This chapter, indeed this book, constitutes an argument to the contrary.

6. Throughout this book, the economic analysis of behavior will be used to exemplify teleological behaviorism. However, it is important to note that many other varieties of teleological behavioral analysis exist. Richard Herrnstein and William Vaughan's (1980) "melioration" and John Gibbon's (1977) "scalar

expectancy" are two examples. So, despite its name, is Peter Killeen's (1992) "behavioral mechanics."

7. Correspondences between cognitive decision theory and teleological behavior theory are further elaborated in Rachlin (1989).

8. However, some psychologists, modern descendants of the gestalt psychologists (see Chapter 1), have applied teleological analysis to purely conscious events. One psychologist of this school is Joseph Rychalk (1988). Rychalk's basic datum is phenomenological. He calls goal-oriented actions "telosponses." Both telosponses and the goals around which they are organized may occur wholly in consciousness, independent of overt behavior; they function like Skinner's operants, but (more like "coverants") wholly internally. Rychalk says (1988, p. 238): "A *telosponse* is the person's taking on (premising) of a meaningful item (image, word, judgmental comparison, etc.) relating to a referent acting as a purpose for the sake of which behavior is then intended. . . . When the individual behaves "for the sake of" this purpose he is telosponding or acting intentionally, although this may be exclusively at the level of understanding and not seen in his overt actions." A telosponse is an item in a person's private phenomenal world. The best observer, actually the only observer, of a telosponse is the person who has it. Thus Rychalk's psychology, although nominally teleological, goes contrary to Donagan's interpretation of Aristotelian philosophy at this point.

9. For a comprehensive cognitive interpretation of Aristotle's psychology wholly in line with modern Aristotelian philosophy, see Robinson (1989). Robinson discusses Aristotle's concepts of sensation, perception, imagination, and thought as *internal* causes of or reasons for overt action. Here is Robinson's (1989, p. 69) interpretation of Aristotle's concept of *imagination:* "The activity in the sensory systems can and does outlast the actual stimulus, and this activity is often so similar to that produced by the stimulus as to be sensibly indistinguishable from it. This [internal] activity or movement is the source of *imagination.*" This view of imagination as the product of an internal representation of a stimulus may well be an accurate reflection of Aristotle's conception. But from a behavioristic viewpoint it contains only part of Aristotle's conception, the part devoted to the internal ("subpersonal") causes ("the source," as Robinson says) of imagination. The final causes of imagination (from a behavioristic viewpoint) must depend on the behavior of a whole organism in its environment and not on internal representations at all. Chapter 5 will present a behavioral reconstruction of Aristotle's final-cause analysis of imagination and other aspects of mental life.

10. A modified version of this chapter was published in *American Psychologist* (Rachlin, 1992).

3

Plato

"What has Plato got to do with behaviorism, really? Other than merely paying one's respects?" I am quoting from an anonymous review of the initial outline of this book. Underlying the remark is, probably, the common view of Plato as a transcendental mentalist and behaviorism as a materialist philosophy.[1] Undoubtedly that comment represents the view of the majority of modern classical philosophers as well as laypeople. But neither Plato nor behaviorism need be viewed in this narrow, polarized way.[2] The historian and critic, J. R. Kantor (1963), sees Platonism as less transcendental than behaviorism. Kantor defines psychology as the study of the interaction or "interbehavior" of a responding organism (taken as a whole) with various "stimulating objects." The interaction occurs in an "event field" that affects the relationship. The psychologist may be an element of this field or may stand outside of it. Just as physics is a science that deals with the interbehavior of objects with one another, psychology is a science that deals with the interbehavior of organisms on the one hand and objects (which may include other organisms) on the other.

According to Kantor, science evolved, not from magic or superstition, but from the interbehavioral observations of everyday life. Underlying and preceding physics as a science, there is a "physics" of everyday life consisting of people's understanding of their material environment. Underlying and preceding psychology as a science, there is a "psychology" of everyday life (a folk psychology) consisting of people's understanding of other organisms (including other people).

Plato is seen by Kantor as fitting into this naturalistic system. Kantor

(1963) says, "What is called for, then, if we want to assess the authentic Plato, is to scale off from his writings the various patinations with which later writers have covered his dialogues . . . specifically we must clear away the interpretation that Plato was concerned with transcendental entities" (p. 99).

Kantor views Plato as, like himself, a realist, albeit a misinterpreted one. There is certainly justification in Plato's work for this view. Plato opposes it to the view (which he attributes to Protagoras and Euthydemus) that the world is a creation of the mind of man. Plato says:

> Things are not relative to individuals, and all things do not equally belong to all at the same moment and always. They must be supposed to have their own proper and permanent essence; they are not in relation to us, or influenced by us, fluctuating according to our fancy, but they are independent and maintain to their own essence the relation prescribed by nature. (*Cratylus*, 386d)

To Protagoras's dictum, "Man is the measure of all things," Plato has Socrates ironically reply:

> In general I am delighted with his statement that what seems to anyone also is, but I am surprised that he did not begin his *Truth* with the words, The measure of all things is the pig, or the baboon, or some sentient creature still more uncouth. There would have been something magnificent in so disdainful an opening, telling us that all the time, while we were admiring him for a wisdom more than mortal, he [Protagoras] was in fact no wiser than a tadpole, to say nothing of any other human being. (*Theatetus*, 161c)

Kantor attempts to show, not only that a behavioristic interpretation of Plato is possible, but also that behaviorism (in the form of Kantor's "interbehaviorism") is the proper context from which to view all Greek thought. He does this via an analysis of the cultural matrix within which Plato and Aristotle wrote. My own aim here is much less ambitious. I make no claims about what Plato really meant. I take the dialogues as they have been translated (already, thereby, accepting previous interpretations to some degree) and attempt to put them into the context of modern behaviorism.

Reconstructing Plato

Having said this much, it is necessary to add that Plato, of all the great philosophers, most easily lends himself to bizarre interpretation. Paul Friedlander (1958, 1964, 1969), in his authoritative commentary on Plato's philosophy, indicates that even Plato's ancient biographers knew this. In one mythical biography: "The dying Plato dreams of himself as a singing swan who flies from tree to tree; no hunter can shoot him down. The Pythagorean Simmias interprets this dream as follows: Everybody would try to grasp Plato's thoughts, but each one would make the interpretation fit his own thinking. [Friedlander adds:] Do we not do so to this day?" (1969, p. 39).

Why are Plato's dialogues so malleable? First, there are the usual ambiguities of translation from one language and culture to another. But, in Plato's case, this is only the beginning of ambiguity. There is the dialogue form itself. The dialogues are dramas and, as Friedlander time and again emphasizes, Plato's philosophy is carried as much by the action and the characters of the participants as by the arguments themselves. Thus discovering Plato's philosophy entails some of the same problems as discovering Shakespeare's philosophy. Above all, there is the character of Socrates. His trial, his defense, and his death color not only the dialogues directly concerned with these events but all of the dialogues, even those in which Socrates does not play a direct part. Other participants, and Socrates himself, frequently refer to his "dangerous" acts. How to live? how to die? are repeatedly asked and when these questions are discussed Socrates' life and death loom always in the background. Furthermore, Socrates is frequently ironic. Sometimes, as in the passage about Protagoras quoted above, the irony is obvious. But sometimes it is not, and one interpreter's irony is another's serious discussion. Finally, in addition to the form of the dialogues and Socrates' irony, there is Plato's frequent use of myth.

Whenever Socrates is made to recite a myth he precedes it by the disclaimer that the myth is a sort of guess, not to be taken literally. Myth in Plato is an effort at persuasion at the limit of rational discourse. According to Friedlander (1964, p. 283), "Myth is not a detour into dreamland, but a call to action." It is a "mistake to use them [myths] as evidence for [any particular] Platonic theory of soul." But if myths are a "call to action," one can still ask, "what action?" and thus open the door to various interpretations and reconstructions.

Plato was quite aware that differences in culture make for differences in understanding. Even within the same culture, understanding is limited by differences in education and ability. At one point in *The Republic*, Socrates says, "You will not be able, dear Glaucon, to follow me further, though on my part there will be no lack of good will. And if I could, I would show you, no longer an image and symbol of my meaning, but the very truth as it appears to me—though whether rightly or not I may not properly affirm" (*Republic* VII, 533a). If Glaucon, Plato's brother and fellow Athenian, cannot be expected to follow, still less can it be expected of us. Part of what is "ideal" in Plato's ideal state, *The Republic*, is the common early experience of the rulers—especially their early exposure to the same sorts of music and myth. "We begin," Plato says, "by telling children fables" (*Republic* II, 377a). This early experience is intended to contribute to harmonious behavior between individuals in adulthood. But since our early exposure to music and myth is so vastly different from Plato's, it seems advisable to make what use we can of his myths, without presuming to understand his true intent.

It may be argued that there can be no behavioral reconstruction of philosophy because behaviorism, even in its most verbal form, is a set of rules for prediction and control of overt behavior. For a behaviorist to engage in argument is to appeal to rationality and thus to concede that rationality is important—or at least real. Such an argument was made recently by the philosopher Arthur Danto (1983): "Scientists themselves, in the activities that define them as such, so conspicuously exemplify the working of intentional systems that if explanatory reference to these begs a question, the question of science itself is begged. These dialectical self-entrapments constitute a paraontological proof for the existence of intentional systems" (p. 359).

But this argument presumes that scientific activity consists of something other than overt behavior. If, by what I do and what I say I cause Danto to change, in a certain way, what he does and what he says, I will be satisfied—however many reservations (forever unexpressed by his overt behavior) he maintains internally. A behavioristic argument itself may be less useful or beneficial than some other argument. But that is not Danto's point.

There is, however, a point of view (it would not be a philosophy) sometimes (wrongly) identified with behaviorism, that could not engage in argument without proving itself wrong. That would be a sort of short-sighted hedonism, a view that regards immediate pleasure as the only legitimate aim of life. In Plato's dialog *Philebus,* the character Philebus takes this view. But he retires quickly from the argument. According to Friedlander (1969, p. 310):

> Philebos represents the "pure" pleasure principle, "unlimited" in its essence. It is not a minor detail either that he is lying down, whereas the participants in the conversation itself are thought of as either standing around or sitting. Philebos has "refused to go on" with the discussion (11 c 8). Pleasure cannot be expected to "give an account" of itself—cannot be expected to because engaging in controversy is not pure pleasure. If it does so engage itself, it enters into a pact with its adversary, reason (*logos*), and in this contest is necessarily defeated. Pure hedonism and dialectics exclude each other. For hedonism is essentially hostile to any dialogue, which means it is non-human. The dream of the hedonist will ultimately reveal itself as the life of an oyster. (21c)

It will become clear to the reader, if it is not clear already, that behaviorism is far from "pure" hedonism of this sort.

The Basic Argument

Plato's philosophy revolves around an ethical point. The first, last, and most essential question is, "How should a person live?" Plato deals with this question by considering life as a series of choices between the merely pleasant and the good. But the very acts of distinguishing between the pleasant and the good are the same acts that are called (in our language) virtuous. Although again I want to emphasize that this is a minority position, I shall try to show that Plato may be read as asserting the identity of ethical understanding and ethical action. In a real (as opposed to ideal) society, knowledge of the good cannot be separated from good behavior. Knowledge of the good is difficult because many (but not all) actions that give us pleasure are not good. Pleasure confuses us. Eventually, the bad person will be unhappy, especially when permitted by social circumstances to live a life of pure pleasure. Good behavior is intrinsically harmonious in the long run while bad behavior is not harmonious in the long run (however pleasant it may be in the short run).

Plato's ideal forms, his view of the soul, his dialectical method and all other aspects of his philosophy revolve around this argument. Let us therefore take up its elements one by one.

Ideal Forms

Gregory Vlastos (1975) begins his book *Plato's Universe* as follows (p. 3):

> In English *cosmos* is a linguistic orphan, a noun without a parent verb. Not so in Greek, which has the active, transitive verb, kosmeo: to set in order, to marshal, to arrange. It is what the military commander does when he arrays men and horses for battle; what a civic official does in preserving the lawful order of a state; what a cook does in putting foodstuffs together to make an appetizing meal; what Odysseus' servants have to do to clean up the gruesome mess in the Palace after the massacre of the suitors. What we get in all of these cases is not just any sort of arranging, but one that strikes the eye or the mind as pleasingly fitting: as setting, or keeping, or putting back, things in their proper order. There is a marked aesthetic component here, . . .

According to Plato each part of the cosmos, including nonliving and even manufactured objects, has an ideal role to play (an ideal function) in the context of the cosmos. This highest role of an object is what Plato calls its *form*. The form of an object exists, for Plato, even in the absence of the material object. Even if there were no actual chairs in the world, for instance, the chair's function (however humble) in the overall scheme of things would still exist. Perhaps the most difficult aspect of Plato's thought for us to grasp is the notion of the existence of forms of objects in the absence of the objects themselves.

The modern concept that comes closest to Plato's concept of form is our notion of an ecological niche—the place in nature occupied by a plant or animal (its natural context). Even if there were no dogs, for example, the ecological niche that dogs occupy could still exist. Forms are *ideal*, in Plato's conception, not because they exist in another world but in the sense that no individual object ever perfectly fulfills its function. If (contrary to fact) it were possible for all dogs to die and for everything else to remain as it is, the gap in the cosmos left by the erstwhile dogs would constitute the form of the dog.

Imagine, as another example, the African plains just as they are in every respect, with their flatness, their lions, their elephants, their trees, but without one particular animal—say, the giraffe. That environment provides a sort of niche into which an animal like a giraffe (with a long neck to reach the highest leaves on the trees and to see the lions from far off) fits. So, even if all giraffes were to die, it would be possible to conceive of a form of animal like the giraffe that could exist in that (now empty) niche.

One difference between our concept of ecological niche and Plato's concept of ideal forms is that *everything* in the cosmos, including chairs, has a form, while only living things have an ecological niche. Otherwise, the ideal forms exist in the same place as do ecological niches: here in the world. (That Plato conceived forms to exist in the real world is why Kantor and Friedlander both deny that Plato was a transcendentalist.)

Imagine a football team without a player at a given position. As long as the particular player is absent that 10-man team provides a niche for an ideal eleventh player. Any *particular* eleventh player can only approximate this ideal. Now imagine each of the 11 players separately removed and you have a concept of an ideal football team. But this ideal team or the ideal form of the team does not exist in heaven; it exists in *its* context—the world of sports—the world out there. A great coach might even be able to conceive of such a team and to mold the real players closer and closer to his vision. To gain such a concept a coach would have to actually be a coach, to open his eyes and see what was in front of him there on the field in the context of what a football team has to do—its function. That would require intelligence. You could not become a good coach by retiring to a monastery, turning your eyes inward, closing them, and looking hard.

More to the point, according to Plato, you could not become a good person by retiring to a monastery, turning your eyes inward, and looking hard. (See Chapter 6 for St. Augustine's version of Neoplatonism, a view that diverges from Plato's on exactly this issue.) To be good, according to Plato, you have to look *outward* at the world and see the ideal forms there, an activity that would be better called "outsight" than "insight." It is not an easy intellectual feat to ignore particulars when you are trying to see abstractions. Particular things (like pleasures) may confuse us, hindering our view of abstractions (like goodness), says Plato.

To take still another example, a recovering alcoholic's most difficult job is to perceive how much he or she is drinking (alcoholism is an abstract concept). Each particular drink is so powerful a stimulus that the total count, the amount of drunkenness, the alcoholism itself, is obscured to the alcoholic (however clear it may be to others).

An important difference between our concept of ecological niche and

Plato's concept of ideal forms is that we see niches as changeable but Plato saw forms as permanent. Although niches last longer than the individual animals or plants that occupy them, they may change with changes in external conditions (for instance, the coming of an ice age). But if everything, including the air, the water, the earth, is included in a vast cosmological order, then no external conditions exist, and niches would not change. Thus, ideal forms are permanent forms. Since, as we will see, Plato identifies reality and goodness with permanence, ideal forms are both real (more real than transitory material objects) and good (in the sense that material objects, including individual human beings, function better as they approach their ideal forms).

Ethics As Fundamental

There can be no serious disagreement with the claim that "How should a person live?" is the crucial question of the dialogues. Whenever this question is addressed the action of the dialog stops, so to speak, and Socrates underlines the point. For instance: "Of all inquiries . . . the noblest is that which concerns . . . what a man should be, and what he should practice and to what extent both when older and when young" (*Gorgias,* 487e). And, just a few pages later: "You see the subject of our discussion . . . and on what subject should even a man of slight intelligence be more serious? . . . namely, what kind of life one should live" (*Gorgias,* 500c).

The Republic, "the best known and generally considered the greatest of the dialogues" (Hamilton, 1961, p. 575), is largely given over to a description of a utopian state. But its major purpose is to guide individual behavior. The problem is that individual behavior cannot be understood outside of its context. In the dialogues prior to *The Republic* (accepting Friedlander's ordering), no firm conclusions can be reached about behavior. Friedlander calls these dialogues "aporetic" (full of doubt). But once an ideal *context* of individual behavior is established, moral behavior is definable. In *The Republic* Plato allows no political analysis to go very far without turning back to the implications of that analysis for individual behavior. This approach is made clear at the very beginning: "Do you think it a small matter that you are attempting to determine and not the entire conduct of life that for each of us would make living most worth while?" (*Republic* I, 344e) and "Is it true that the just have a better life than the unjust and are happier? . . . It appears even now that they are, I think from what has already been said. But all the same we must examine it more carefully. For it is no ordinary matter that we are discussing, but the right conduct of life" (*Republic* I, 352e). And, near the end: "Our inquiry concerns the greatest of all things, the good life or the bad life" (*Republic* IX, 578c). *The Republic* takes its place among the dialogues as an essential step toward the goal of describing how a person should live. It provides an ideal context (a form) for individual behavior.

Living As Choosing

The living of the good life is a matter of making intelligent choices under conditions in which choices are constrained. In book X, the last book of *The Republic*, Socrates says:

> It should be our main concern that each of us, neglecting all other studies should seek and study this thing—if in any way he may be able to learn of and discover the man who will give him the ability and the knowledge to distinguish the life that is good from that which is bad, and *always and everywhere to choose the best that conditions allow* . . . so that he may make a reasoned choice between the better and the worse life. (*Republic* X, 618c; italics mine)

An ideal state would be arranged so that people would naturally live good lives. The ideal state would provide constraints and rewards so as to bring people's short-term interest (their "pleasure") into line with their long-term interest (their "happiness"). As it is, behavior that results in pleasure often conflicts with behavior that results in happiness (good behavior) so people are forced to choose.

The reason why so many intelligent people find it so difficult to make choices is that, like the coach without a quarterback on his football team, as long as there is no *real* quarterback on the team there exists an abstract quarterback, an *ideal* quarterback, the perfect Platonic form of the quarterback, a niche or gap that could, only in theory, be perfectly filled. But when it comes to particulars, just as there is no particular quarterback that can perfectly fill the gap, so there is no particular husband, no particular wife, no particular child, no particular anything, that can perfectly fulfill the function we have laid out for it.

As long as the coach does not fill the quarterback slot on his team, as long as we do not get married, as long as we do not have a particular child, as long as we do not take a particular job (and as long as we do not write a particular book), we in a sense retain the abstract, the ideal, versions of these things. As soon as we accept particulars, though, no matter how good they are, we lose the generalities that we have been coddling.

Faced with a choice between two paths, to *take* neither particular path is, in some Platonic sense, to *have* both paths in the abstract. To take one path is to give up the other. The more intelligent we are, the better our vision of these perfect abstractions (our "outsight") and therefore, according to Plato, the more reluctant we will be to give them up–to make the particular decisions that life requires.

The problem is that eventually life has to be lived in particular terms. The coach cannot play football with a team without a real quarterback (however far from ideal), and we require husbands, wives, children, jobs, in order to live. We cannot construct a life out of gaps and abstractions; we have to construct our lives out of particular events. Intelligent people therefore cannot rest with their view of life as an abstraction but, as Plato himself said and as Aristotle emphasized, must go to work, actually live their lives, and simultaneously try to approach abstract happiness (wisdom, according to Plato) as closely as possible.

The Pleasant and the Good

If the object of philosophy is to discover how to live, and living often consists of choices between the pleasant and the good, then we need to ask what, exactly, is the difference between them? No subject is analyzed more extensively in the dialogues than this one. In one of the earliest dialogues, *Protagoras,* Plato seems to take the view that there is no difference between the pleasant and the good except in terms of delay:

> Pleasant, painful, good and evil . . . call them by only two names—first of all good and evil and only at a different stage pleasure and pain. . . . So like an expert in weighing, put the pleasures and the pains together, set both the near and the distant in the balance, and say which is the greater quantity. *Protagoras,* 355b, 356b)

This may be seen in terms of the standard behavioral model of self-control illustrated in Figure 3.1. The larger but more distant benefit corresponds

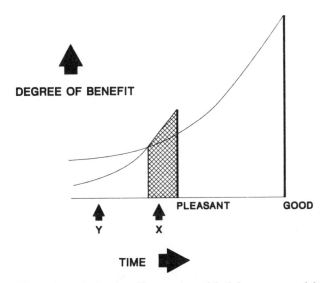

Figure 3.1 A simple self-control model. A larger, more delayed reward (GOOD) and a smaller, more immediate reward (PLEASANT) are both discounted with delay by the same (hyperbolic) function. When both rewards are relatively distant (point Y) the larger is more valued but as time passes, the present value of the smaller reward grows faster than that of the larger until when the smaller reward is immanent (point X) its present value exceeds that of the larger. At point X (and generally during the period indicated by the cross-hatching) the pleasant reward constitutes a "temptation."

to the good, and the smaller but near benefit corresponds to the pleasure. The gradients (thin lines) represent, in Plato's terms, how big the benefits "seem." In the cross-hatched area the smaller benefit subtends a higher gradient. Thus, at time X, the smaller benefit "seems" larger. For Plato, "the same magnitudes seem greater to the eye from near at hand than they do from a distance" (*Protagoras,* 356c). At a greater temporal distance (time Y in Figure 3.1), the model says that the ratio of the smaller and larger benefits seems closer to its true value. When benefits are close to us, however (at time X), we cannot trust our perceptions. At that point, "appearances lead us astray" (*Protagoras,* 356d) and the only way to secure our happiness is by measurement and calculation of the true benefits. In this sense, knowledge obtained through measurement and calculation is superior to perception. It is a theme that runs through all of the dialogues that perception (for reasons similar to those given above) is a poor guide to reality. Furthermore, once a person has made the necessary calculations and knows which alternative is better, he cannot choose the worse alternative. If he does he has *by definition* calculated wrongly. Thus, "no one ever willingly goes to meet evil" (*Protagoras,* 358c). A person's knowledge, for Plato, *necessarily* corresponds to the choices he makes. (I shall return to this issue later.) The fact that one can measure and compare the pleasant and the good does not imply that they must be the same thing. In *Protagoras,* Socrates assumes an uncharacteristically hedonistic position, perhaps to trick Protagoras into agreeing that there is a general concept of good independent of each person's individual desires.

Behaviorists have been criticized for viewing all human goals as in some sense comparable (e.g., Savin, 1980; Schwartz & Lacey, 1982) but, as Plato shows, life demands comparisons between qualitatively different human goals because it demands choices between them. Even in later dialogues, after Plato has distinguished in a more sophisticated way between the pleasant and the good, and has shown their relationship to involve vast qualitative differences, he still persists in viewing the pleasant and the good as measurable and quantitatively comparable. For instance, in *The Republic* (IX, 587c, d, e), he makes a comparison between the happiness of the philosopher-king and that of the tyrant. Between the king and the tyrant (in Plato's political hierarchy) stand the oligarch and the democrat, and thus the difference lies in the nature of their souls (3 units) compounded by their relationship to others in the state ($3^2 = 9$ units) and compounded again by the relation of their states to other states ($9^3 = 729$ units). Thus the philosopher is 729 times as happy as the tyrant. Obviously, Plato is not wholly serious here but still, according to Friedlander (1969), "everything that precedes aims at this final reckoning" (p. 119). Similarly, the fact that behavioristic measurement techniques work easiest in the laboratory with clearly defined rewards does not mean that these techniques cannot be applied to choices among complex human goals. "We observe," Friedlander (1964) says in his analysis of *Euthydemus,* "how Plato is at pains to divest human happiness of any accident and ground it in an act of choice" (p. 186). Friedlander could have been coining a motto for modern behavioral research.

Plato's common dimension of comparison between pleasure and good is *temporal*—how long-lasting are the benefits of each. The calculated ratio, Plato says, "is a true number and pertinent to the lives of men if days and nights and months and years apply to them" (*Republic* IX, 588a). This contrasts with the view that choice between pleasure and goodness is based on a comparison among (internal) emotional states. About this possibility, Plato says:

> From the moral standpoint it is not the right method to exchange one degree of pleasure or pain or fear for another, like coins of different values. . . . A system of morality which is based on relative emotional values is a mere illusion, a thoroughly vulgar conception which has nothing sound in it and nothing true. (*Phaedo,* 69a)

The measurable difference between the good and the pleasurable, then, lies in their relative durations. This temporal difference, however, is the inevitable outcome of a structural difference between them. In the dialogue *Gorgias* (classified by Friedlander, not among the early, aporetic dialogues, but among a middle series, grouped under the heading, "The Logos Takes a Stand," and seen by him as a sort of warm-up for *The Republic*), Socrates argues that the pleasure-versus-pain dichotomy cannot be the same as the good-versus-evil dichotomy. Good is necessarily the opposite of evil; it is what we *mean* by the opposite of evil. But pleasure is not what we mean by the opposite of pain. In fact, pleasure and pain often go together. Pleasures like eating and drinking, require an initial deprivation and thus only involve a reduction of that deprivation to some sort of zero point (*Gorgias,* 497).

In behavioristic terms, Plato is saying that at least some pleasures are negative rather than positive reinforcers. Such pleasures are always, according to Plato, mixed with pain. When there is no more pain (no more deprivation), there is also no more pleasure. If good is the opposite of evil and pleasure is *not* the opposite of pain, the good-evil dichotomy cannot be the same as the pleasure-pain dichotomy.

Another difference between pleasure and good is that many activities that are unquestionably pleasurable have painful consequences where the ensuing pain greatly outweighs the pleasure. Now (we shall come back to this in a moment) an act and its consequences taken together may be evaluated as a single entity. A pleasurable act outweighed by painful consequences (overeating, for instance) may on the whole be bad. Thus, we can have bad pleasures as well as good pleasures and good pains (like having a tooth extracted) as well as bad pains (*Gorgias,* 499). We often have a choice between pleasure (like overeating) and good (like being healthy) and between painfulness (like taking medicine) and harm (like being unhealthy) and we should generally choose the good over the pleasurable and the painful over the harmful. But if pleasures may sometimes be bad (opposite to good), they cannot always be good. Thus, the good cannot be defined in terms of the pleasant. It must be defined in some other way. How then?

Goodness and Reality

According to Friedlander (1958), "the *Idea* of the Good gives being and order to the objects of the world of being" (p. 63). Friedlander is saying that, for Plato, ontology (the theory of being or reality) is at the service of ethics. It may be hard for us to imagine (in other than religious terms) how goodness can possibly stand above reality (and Friedlander's reading of this conception into Plato is at the very least controversial). For us "reality" is a bottom-line word. (If reality is good, fine. If reality is, as we sometimes suspect, not good, then we ought to face it.) Perhaps a better way of understanding Plato's conception is to ask what is the function or purpose of the order and harmony that not only Plato but virtually all Greek philosophers believed existed in the cosmos (Vlastos, 1975). In Plato's view (as we shall soon see) the true nature of a thing (its reality) is determined by its function. (The true nature of a chair, for instance, is its sittableness.) The function of the order in the cosmos is, according to Plato, to serve as a guide to our behavior—to make us good.[3] It is in this sense that the reality of the cosmos (the order we find there) is subservient to the good. But this sense is the most important sense because for Plato the purpose of philosophy itself is not to understand the world but to help us to live a good life. Thus, we call some things "real" and others "not real" because making this distinction will help us live a good life. We do not adjust our concept of what is good to what is real; we adjust our concept of what is real to what is good.

For Plato, an object becomes better by rising from a lower level to a higher level of reality (by approaching an ideal form). What makes a high level high is its relative *permanence*. If an object is continuously changing, it cannot be very much worth having. "What great thing . . . could there be in a little time?" Socrates asks (*Republic* X, 608c). By implication, none. A brief event might give pleasure, but *because* of its changeability, cannot be a great thing, or even much of a thing at all.

Just as the main characteristic of goodness as opposed to pleasure is that goodness is *lasting*, the main characteristic of reality as opposed to illusion is that reality is *lasting*. But material objects themselves are always changing; what lasts are relationships or patterns between objects. In the case of the relation between people and objects the thing that lasts is the *function* of the objects:

> Do not the excellence, the beauty, the rightness of every implement, living thing and action refer solely to the use for which each is made or by nature adapted? (*Republic* X, 610d)

The person who knows a chair in its truest sense is its user. The carpenter must grasp the idea of a chair, not as a maker, but as a user of chairs. The chair's function both precedes and follows the existence of any individual chair. This notion of the importance of function makes Plato (in modern terms) an antireductionist (a molarist). In a famous passage from the *Phaedo,* Socrates, about to die, considers molecular and molar explanations for his position:

> The reason why I am lying here now is [not] that my body is composed of bones and sinews and that the bones are rigid and separated at the joints, but the sinews are capable of contraction and relaxation, and form an envelope for the bones with the help of the flesh and skin, the latter holding all together, and since the bones move freely in their joints the sinews by relaxing and contracting enable me somehow to bend my limbs. . . . The real reasons . . . are that since Athens has thought it better to condemn me . . . I for my part have thought it better to sit here, and more right to stay and submit to whatever penalty she orders. (*Phaedo,* 98c, d)

The critical relationship underlying Plato's concept of reality is a molar functional relationship; in behavioristic terms, an instrumental relationship. The way in which such relationships bring about an elevation in reality is the subject of *The Symposium,* a dialogue consisting mostly of a series of speeches about love. In this dialogue, the other participants speak eloquently in praise of love as a god. But Socrates, as always in the dialogues, wants to understand the general principles underlying the concept. He expands the discussion as follows:

> "Love, that all-beguiling power," includes every kind of longing for happiness and for the good. Yet those of us who are subject to this longing in the various fields of business, athletics, philosophy and so on, are never said to be in love, and are never known as lovers, while the man who devotes himself to what is only one of Love's many activities is given the name that should apply to all the rest as well. (*Symposium,* 305d)

Plato expands the notion of love to cover all (of what we would consider to be) instrumental behavior. An instrumental act is one in which a lesser-valued activity is performed for the sake of a higher-valued activity (see Chapter 2). The two activities together form a pattern that is still higher valued, higher than either of its components alone. As Friedlander (1969) puts it, "In love that which is neither good nor bad but rather in between strives upward toward perfection" (p. 21).

Figure 3.2 illustrates a corresponding behavioral model (which the interested reader will find applied in some detail to animal behavior in Staddon and Motheral, 1978, and to the play of children in Rachlin and Burkhard, 1978). Graphically, a diagonal plane (making an equilateral triangle as it intersects the three planes defining Cartesian space) intersects concentric spheres. Point *d,* at the center, represents a maximum. Let us

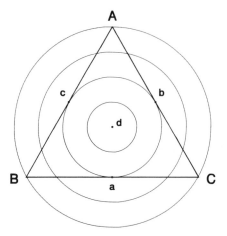

Figure 3.2   Values represented by circular contours with apex at point *d*. Value
decreases with distance from *d*. Mixtures of *A*, *B*, and *C* are represented by
points on the sides and inside of the triangle *ABC*. The line *AbC* is a one-
dimensional projection of the two-dimensional triangle *ABC* in a sense
analogous to Plato's cave allegory in which two-dimensional particulars are
projections of three-dimensional reality.

call it the best combination that can be achieved consisting of the entities
A, *B*, and *C*, which may be, in Plato's terms, "implements, living things, or
actions." Point *d*, the maximum, stands, as it were, on the top of a hill. The
concentric circles are (like onion rings) a two-dimensional slice of a three-
dimensional spherical utility function $[U = (A - x)^2 + (B - y)^2 + (C - z)^2]$. Like altitude contours on a map, the circles represent, as they
increase in diameter, ever-decreasing levels of value, ever further distances
from point *d*. In the diagram a point closer to *d* (higher on the hill) is more
valuable than a point further from *d*. The triangle is a way to represent the
constraints imposed by three entities (actions, for example) in two dimensions. (Point *A* represents act *A* alone, point *B* represents act *B* alone, and
point *C* represents act *C* alone. Points on side *A-B* (such as point *c*)
represent combinations of acts *A* and *B* (with no amount of *C*), and points
on the other sides represent corresponding twofold combinations. Points
within the triangle represent combinations of acts *A*, *B*, and *C*. The trian-
gular form constrains the combinations so that if one entity is increased
another must give way $(A + B + C = T$, where *T* is a constant). For
instance, the triangle could represent three activities that fill (and are
constrained within) a fixed period of time.[4]
      The property of this representation that makes it mirror Plato's theory
of value is the fact that points *A*, *B*, and *C* alone $(A = T; B = T; C = T)$
are relatively far from the maximum, point *d*. Thus, to stay within a single
dimension of the triangle is to stay at a relatively low-valued point. The
only way to increase value is by mixing activities. Some mixtures would be

better than others. For instance, the mixture of *A* and *B* at point *c* would be closest to *d* of all possible mixtures of *A* and *B*. By going from one dimension to mixtures of two dimensions one could move from points *A*, *B*, or *C* to points *a*, *b*, or *c* and come closer to point *d*. But only by mixing all three entities could point *d* actually be reached.

The measure of good in Plato is always how abstract something is. Lower levels of abstraction are seen by Plato as projections of higher levels, analogous to geometric projections—as triangle *A-B-C* may be projected onto line *A-B* and line *A-B* onto point *A*.

There is no reason, except convenience of representation, to limit the utility function to three dimensions. Just as expansion from one to two dimensions and two to three dimensions allows value to increase, so expansion to more dimensions would allow value to increase still further. As in Plato's conception, any level actually achieved, no matter how high, is provisional. It is always possible to imagine some new dimension allowing a more complex pattern of behavior that would be still better. For Plato, the first dimension is the material world, the world of immediate perception. Subsequent dimensions are relationships, patterns, or proportions of things to one another. An increase of the number of dimensions is an increase in the level of abstraction.

Knowledge

In *The Symposium* (210) Plato describes an ascent from love of "one individual body" to "every lovely body" to "beauties of the soul" to "beauty of laws and institutions" to "beauty of every kind of knowledge" to "the very soul of beauty." Love, in Plato's conception, is not a god (which would necessarily be perfect) but a "spirit" or a force (like the concept of force in modern physics) that stands for our tendency to move upward to higher levels. In one of the later dialogues, speaking of happiness as the mixture of pleasure and intelligence, Plato says:

> Why it's just as if we were supplying drinks, with two fountains at our disposal; one would be of honey, standing for pleasure, the other standing for intelligence, a sobering, unintoxicating fountain of plain salubrious water. We must go to work and make a really good mixture. (*Philebus*, 61c)

It is the job of the philosopher to study the blending and combination of things. The philosopher is the expert who knows best "how to distinguish, kind by kind, in what ways the several kinds can or cannot combine" (*Sophist*, 253d).

The knowledge of the philosopher is not a knowledge of material things but of relationships among things. (These relationships do not simply occur in the head of the philosopher. Plato sees them as actually out there in the real world. They have their own "proper and permanent essence.") The more abstract a relationship is (the more dimensions it has—the more context it takes into account), the better it can be known; hence, the more *real* the objects of knowledge. Thus, Plato's concept of reality arises from his concept of goodness. The idea of the good stands to the material world as the sun stands to the visible world. As the sun gives light to visible objects and enables us to see them, so the idea of the good gives existence or reality to the objects of knowledge and enables us to understand them (*Republic* VI, 508, 509).

In *The Republic* (V, 476) Plato distinguishes between knowledge and opinion. A person who knows is "awake" while a person who just has opinions is "in a dream." The difference between objects of knowledge (reality) and objects of opinion is a difference of context. Dreams lack context, that is, they lack consistent relationships with other things. Thus context is the mark of reality. Reality is separated from the good, which incorporates all context, on the one hand, and from dreams, which have no context, on the other.

People who have true belief, for Plato, are like "blind men who go the right way" (*Republic*, VI, 506c). They choose the good by accident, because it happens to coincide with the pleasurable. But for knowledge, we need "true belief" plus "an account" (*Theatatus*, 201). The account puts the belief in its context and relates it to other things. It allows long-term consequences to influence choice. Plato realizes that a single object may rationally be placed in a multitude of different contexts. That is why, ultimately, what forms the proper context of an object will be an aesthetic judgment. At this point, the "good has taken refuge in the character of the beautiful, for the qualities of measure and proportion invariably . . . constitute beauty and excellence" (*Philebus*, 64e). The difference between insane behavior and normal behavior would not be that one has no context and the other does, but that the context of insane behavior does not conform to certain aesthetic criteria. In Plato's terms, the soul of the insane person (or the criminal) is disharmonious. That is why it is so essential, for Plato, that music (along with gymnastics) be the foundation of education and why art, in an ideal state, be so tightly controlled—even to the exclusion of great poetry—if it does not foster a common sense among the population of what is harmonious and what is not. "The modes of music," Plato says, "are never disturbed without unsettling of the most fundamental political and social conventions" (*Republic*, IV, 424c). The answer to the question, How should one live? is to choose a balanced or well-mixed or harmonious life: "always to choose—the life that is seated in the mean and shun the excess in either direction, both in this world so far as may be and in all the life to come, . . . this is the greatest happiness for man" (*Republic*, X, 619b). This balance within the individual is the counterpart of that within the state. In the state, this balance (justice) is achieved by each person performing his or her proper function: "the having and doing of one's own." In the individual the balance is of the soul with each part of the soul playing its own role in relation to the rest.

The Soul

There is some ambiguity in Plato's concept of soul. On one hand, it may be thought of as wholly interior to the person, its parts interacting as the parts of a machine, harmoniously or disharmoniously as the soul is good or bad. On the other hand, the parts of the soul may be seen as a set of functional attributes of overt choice behavior. The former interpretation is a cognitive one (as cognitive psychology was defined in Chapters 1 and 2). It is certainly the easiest interpretation for us to make, relative to our own culture and our own understanding of what "soul" ought to mean. The latter interpretation is behavioral. Is it legitimate?

In a myth in the dialogue, *Phaedrus,* Socrates speaks of the soul as divided "into three parts, two being like steeds and the third like a charioteer" (*Phaedrus,* 253c). The charioteer, by being raised up, can see distant goals but the two steeds cannot. Of the steeds, the one corresponding to spirit is obedient; "the other is crooked of frame, a massive jumble of [a] creature, with thick short neck, snub nose, black skin, and gray eyes; hot-blooded, consorting with wantonness and vainglory; shaggy of ear, deaf, and hard to control with whip and goad" (*Phaedrus,* 253e). The best soul is one in which all three parts work together in harmony under the charioteer's control. One may interpret the three parts of the soul in religious terms or Freudian terms or cognitive terms. But it is also possible to see them in behavioral terms.

In *The Republic* (IV, 436) the context of the discussion of the soul is the question, How should a person allocate labor and time among various activities? In this context it is possible to see each part of the soul as a principle of choice attuned to different outcomes and to see the balance between the parts as a balance of choice outcomes. "The three parts [of the soul] have three kinds of pleasure, one peculiar to each, and similarly three appetites and controls" (*Republic* IX, 580d). The charioteer wants "wisdom," the good horse wants "victory," and the bad horse wants "gain." When the soul, as a unit,

> accepts the guidance of the wisdom-loving part and is not filled with inner dissension, the result for each part is that it in all other respects keeps to its own task, and is just, and likewise that each enjoys its own proper pleasures and the best pleasures and, so far as such a thing is possible, the truest. (*Republic* IX, 587a)

The sense in which these pleasures or goals are parts of the soul, is not as mechanisms like the carburetor, steering wheel, and engine are parts of a car, but as "ruling principles" (*Republic* IX, 581c). In other words, they are functions or ends—outcomes. A person who, given a series of opportunities to choose, makes a series of choices that conforms to a certain pattern may be said to have a soul representative of that pattern. If a person's *actual* choices are best explained in terms of long-term benefits, that person may be said to have a soul dominated by the wisdom-loving part. And, if all of life itself is viewed as a series of choices (as Plato does seem to view it), then the person's soul is that person's pattern of overt behavior and not anything that occurs inside the person. Where do these patterns come from? From the universe (the cosmos) is Plato's answer:

SOCRATES: Shall we not admit that the body belonging to us has a soul?

PROTARCHUS: Plainly we shall.

SOCRATES: And where, Protarchus my friend, could it have got it from if the body of the universe, which has elements the same as our own though still fairer in every respect, were not in fact possessed of a soul?"

(*Philebus*, 30a)

In other words, the soul may be seen as a kind of matching of the form of individual behavior to the underlying form of the universe. Philosophers, in constructing the ideal state (after having wiped it clean of its faults), will look to justice, beauty, sobriety, and the like "as they are in the nature of things" (*Republic* VI, 501b) and then try to produce them in mankind. On the level of individual morality, a person should match his own behavior to the widest possible aspect of the nature of things: to reality (*Republic* VI, 490b). Then we will have, in Friedlander's terms, an "ordered system, or cosmos, as it were, in rightful control over a body endowed with soul" (Friedlander, 1969, p. 348). Harmony in the soul need not, therefore, be viewed as a harmony within a person but a harmony of actions of the person, taken as a whole, with the rest of the world: "a man 'equilibrated' and 'assimilated' to virtue's self perfectly so far as may be, *in word and deed* [behavior] and holding rule in a city of like quality" is how Plato describes a philosopher-king (*Republic* VI, 499a; italics mine).

Private Knowledge

The test for the behavioral view is whether it is possible to have private knowledge of your own soul—knowledge that may never be revealed in behavior. If it is possible to know your own soul without actually making overt choices, then the soul cannot be *identical* with a pattern of such choices and the behaviorist view must be wrong.

Is it possible, according to Plato, for a person to perform *intrinsically* private cognitive actions? In one sense of "private," we all perform private actions. As I am writing these words I am alone in a room and, I assume, completely unobserved. In that sense, my writing is a private act. But are there acts, meaningful cognitive acts, that I could perform, unobserved, surrounded by people, all closely attending to my overt behavior for long periods of time? From the teleological behaviorist view (as Chapters 1 and

2 define it) such intrinsically private acts would be impossible. [It is, of course, possible to talk to yourself or to make unobserved neuromuscular movements but these acts, the behaviorist claims, are (like dreams for Plato) not meaningful: they are not connected to other acts in meaningful ways; they have neither the contextual breadth nor the functional significance that Plato attributes to reality.]

What does Plato say about private cognitive events? First, he emphasizes that "we cannot want what is bad for us" (*Gorgias,* 486e) except out of ignorance. Our knowledge of what will benefit us in the long run is not separable from our actual choice of what will benefit us in the long run. There is a *necessary* connection between what we know and what we desire on the one hand and between what we desire and what we choose on the other. Knowledge necessarily implies one sort of behavior while ignorance necessarily implies another sort; we cannot have private, unexpressed knowledge:

> The sound-minded man would do his duty by gods and men for he would not be sound of mind if he did what was unfitting. That must *necessarily* be so. And doing his duty by men, he would be acting justly, and doing it by the gods, piously, and the doer of just and pious deeds *must* be just and pious. That is so. (*Gorgias,* 507b; italics mine)

And, later, "does not doing just acts engender justice and unjust injustice?" The reply is: "Of *necessity*" (*Republic* IV, 444d; italics mine).

In these passages Plato refers to a necessary connection between mental states and behavior. Nowadays we tend to think of a necessary connection between two things as the operation of a physical law based on efficient causation between logically distinct entities. But it is important to keep in mind that during Plato's time physical laws had not yet become paradigm cases of causal relations. For Plato, it is much more likely that a necessary connection is a logical one. (For Aristotle, this was explicit, as we will see in Chapter 4). Thus, when Plato says that a doer of just and pious deeds "must be" just and pious, he means that there is no logical difference between a just and pious person and a person who acts justly and piously. This, of course, is exactly what the behaviorist says.[5]

Plato's view of private knowledge is perhaps best expressed in the famous allegory of the cave in *The Republic.* Having just previously distinguished among levels of reality (with a model something like that of Figure 3.2), Plato now considers "our nature in respect of education and its lack" (*Republic* VII, 514a). (Note the word 'education' rather than 'self-awareness,' 'insight,' 'self-knowledge,' etc.) He imagines prisoners chained in a cave and forced to view all actions of themselves and each other as shadows (two-dimensional projections) on a wall. Then he imagines what it would be like for a person to be freed and brought out of the cave and into the sunlight of the three-dimensional world. He describes the initial blindness and the subsequent enlightenment that the freed prisoner would experience. Then, in the allegory, the same person goes back to the cave where it takes him a while to readapt. He tries to explain to the prisoners what the shadows really represent. Of course, the unfreed prisoners cannot understand. They see him as a fool at best and, at worst, a grave danger (a reference to Socrates' fate). In interpreting this allegory, the journey into the sunlight has been seen, mentalistically, as an *interior* journey, leading to truer internal reflection, hence truer introspection (see the discussion of St. Augustine in Chapter 6). Of course, an allegory invites all sorts of interpretation. However, the critical point is that the prisoners in the cave do not see "anything *of themselves* or of one another" (*Republic* VII, 515a; italics mine). And the man who journeys out of the cave and becomes enlightened learns about his own reality at the same time as he learns about the reality of other people and of the world. By analogy, all of us learn about ourselves in the same way as we learn about other people and about the world. If we could not understand other people or the world through introspection, we could not understand ourselves that way either.

The mentalist might agree up to this point. Indeed, the standard transcendental interpretation of Plato is that we do learn about ourselves, other people, and the rest of the world in one way: by internal reflection. One justification for this interpretation comes from Plato's view of learning as recollection and thought as reflection.

Recollection and Reflection

Both recollection and reflection may be held to be internal processes. But it is far from clear that Plato saw them as such. Plato's conception of the learning process bears a strong resemblance to Skinner's (1938) conception of shaping of behavior (and vice versa, of course). In a passage in the dialogue *Meno* Socrates teaches an intelligent but ignorant slave boy some geometrical facts by a method much like modern programmed instruction. The boy is led through a series of stages where all he does is answer questions by a "yes" or "no" or "double" or "half" or "it must" or "it looks like it" and at the end is found to know geometrical relationships. The boy "already knew" the answers in the same sense that a pigeon "already knows" how to peck a key or a rat "already knows" how to press a lever. Behavior, to be reinforced (in Skinnerian terminology), must first occur at some "operant level" (some preexperimental baseline). Plato's process of recollection may be seen as nothing but the differentiation by the environment of behavior already in the boy's repertory. As Friedlander (1964, p. 282) puts it, in reference to another dialogue, "Socrates . . . shows that it is the pupil, not the teacher, the respondent, not the questioner, who 'brings forth what is said.' " The teacher's role is to ask questions and (in Skinner's terms) reinforce the correct answers. The requirements for a good teacher are specified in *The Republic* as follows: "There might be an art, an art of the speediest and most effective shifting or conversion of the soul, not an art of producing vision in it, but on the assumption that it already possesses vision but does not rightly direct it and does not look where it should, an art of bringing this about" (*Republic* VII, 518d). The reason that Plato emphasizes the fact that people *already know* what they are about to learn is that he objects to what in modern terms would be called "the storehouse theory of knowledge." Education is "not like inserting vision into blind eyes." It is not, in fact, an *internal* change at all. It is a change of behavior in the *whole body*. "The true analogy for this indwelling power of the soul and the instrument whereby each of us apprehends is that of an eye that could not be converted to the light from the darkness except by turning the *whole body*" (*Republic* VI, 518c; italics mine).[6]

It may be argued that the point of Plato's thought experiment in the *Meno* was that the boy already *knew* the principles of geometry, not that he could *do* the problem. Knowledge in some sense is infinite in scope but doing problems is finite. Here I hope the reader will recall the molar character of teleological behaviorism. To reduce the issue to its most basic terms, *probability* of response is an abstract conception. There are an infinite number of specific actions that compose the same probability of response. To say that a rat presses a lever with a given probability is thus *already* to abstract a general quality from a specific behavioral pattern, to classify the rat's behavior and in some sense to predict what sorts of specific things the rat will do in the (relatively near) future. The relation of the boy's *knowledge* in the Meno myth to the boy's actual behavior is indeed the relation of something infinite and abstract to something particular and concrete. But so is the relation of a probability of response as a molar characteristic of behavior to what an animal actually does. Thus, "recollection" (learning) corresponds to molar behavior. It need not be an internal process at all. How about "reflection" (thought)?

According to the behaviorist, we learn about reality through making choices and observing their consequences. (Higher levels of reality correspond to more complex and long-term consequences.) In the case of our own behavior our only advantage is that we usually have more information about ourselves than about others (more data, but by no means better data). When Plato refers to "reflection" he may be understood as referring to the process by which the environment and behavior reciprocally influence each other (Kantor's "interbehavior"). A person painting a picture of a scene would be reflecting the scene in this sense. When Plato writes, "Knowledge does not reside in the impressions but in our reflection upon them. It is there, seemingly, and not in the impressions that it is possible to grasp existence and truth" (*Theatetus*, 186d), he is not (the behaviorist would say) comparing the earlier stage of an internal process to a later stage (as modern cognitive psychology would have it) but is comparing a simple, one-way, process (from environment to organism) to a more complex interaction (from environment to organism and back).

To summarize, the allegory of the cave shows, among other things, that we learn about ourselves in the same way we learn about other people and the world. The mentalist claims that this learning comes about through introspection. The behaviorist claims that this learning comes about through interaction of our whole bodies with the environment. The Platonic concepts of recollection and reflection that may seem to indicate that Plato was a mentalist actually have coherent behavioral interpretations consistent with his use of these terms.

Dialectics

It may be argued (as perhaps Danto might argue) that reasoning is an intrinsically internal process and cannot be accounted for in behavioral terms. But whatever one thinks about reasoning in general, Plato's preferred form of reasoning—the dialectical method—is certainly not internal. In fact, Socrates claims that it cannot be practiced alone but is really a social activity. To do it properly you need a conversation where a give-and-take is possible. In dialectic (the origin of deductive reasoning as codified by Aristotle), a disagreement is resolved by first discovering grounds of agreement (the premises) and then drawing logical conclusions from them (showing that the conclusions are subcategories of the premises) so as to confirm one side of the original disagreement and refute the other. In the dialogues, this process is one-sided. Socrates usually provides the disagreement, proposes the premises, and clears the logical path to the conclusions. Nevertheless, it takes two parties to have a disagreement in the first place, an agreement in the second and a being-convinced (or not-being-convinced) in the third.

Relative to discussion, even writing loses spontaneity and hence is inferior. The conversation in the *Meno* between the slave boy and Socrates, in which the slave boy already "knows" the answers to Socrates' questions and Socrates acts only as a "midwife," to help bring out the answers, "is the simplest possible formulation of the dialectical method as practiced by Plato's Socrates" (Friedlander, 1964, p. 282).

The dialectical process is social and verbal but the object of dialectics is not just to compel agreement. As Friedlander (1964) says, "words are envisaged as instruments after the model of tools in the crafts" (p. 200). In the later dialogues the dialectical method consists of the practice of functional categorization. A verbal category is taken and subdivided into parts and then subdivided again and again. The object of the exercise is to try to discover what are the basic *forms*. The true forms, for Plato, are not linguistic categories but *functional* ones. In the dialogue *Cratylus,* mostly given over to speculation on the origin of words, Plato compares the way we choose names to the way the weaver chooses a shuttle: "Whatever is the shuttle best adapted to each kind of work, that ought to be the form the maker produces in each case" (389c).

In *The Republic* (IV, 453e) Plato illustrates the difference between *dialectic* which has functional relevance, and *eristic*, which is an arbitrary categorization done only for the sake of argument. You could divide mankind first into men and women, and then assign occupations to each sex. But because both men and women are fit for almost all occupations, this sort of division is not functional. It is like dividing mankind into hairy and bald and then assigning occupations to each. The mistake in both cases is that the division does not take account of the "aim intended." Ultimately, the aim intended is to use language to guide behavior. Thus, for Plato, dialectic is an overt social activity by which words are used to provide functional definitions of forms, *not* an intrinsically private activity that could possibly occur inside a person while he was sitting still or doing something else.

Happiness

It is not only the concepts of justice, piety, knowledge, and reasoning in Plato's philosophy that easily lend themselves to behavioral interpretation but also the ultimate human goal: happiness. Near the end of *The Republic* Socrates discusses how to evaluate the happiness of the tyrant. He rejects the validity of the tyrant's introspection. Instead, the best—the *ideal*—way to evaluate the tyrant's happiness is described as follows:

> The man to whom we ought to listen [about the tyrant's happiness] is he who has this capacity of judgment [not to be overawed] and who has lived under the same roof with a tyrant and has witnessed his conduct in his own home and observed in person his dealings with his intimates in each instance where he would best be seen stripped of his vesture of tragedy, and who had likewise observed his behavior in the hazards of his public life. (*Republic* IX, 577b)

Failing this, the only way to gauge the tyrant's happiness is to "recall the general likeness between the city and the man and then observe in turn what happens to each of them" (*Republic* IX, 577c).

That the good life and the happiness it entails in the soul are one and the same thing—a harmonious pattern of overt behavior (a series of choices)—is stated quite specifically in the dialogue Friedlander places immediately after *The Republic*, the *Theatetus*:

> There are two patterns, my friend, in the unchangeable nature of things, one of divine happiness, the other of godless misery—a truth to which their [unjust peoples'] folly makes them utterly blind, unaware that in doing injustice they are growing less like one of these patterns and more like the other. The penalty they pay is the life they lead, answering to the pattern they resemble. (*Theatetus*, 176e–177a)

This is part of an argument against the notion that "man is the measure of all things." According to Plato it takes an expert, the philosopher, to match his behavior to the more complex patterns of things. Socrates says, with the usual heavy-handed irony with which he always discusses Protagoras:

> We may quite reasonably put it to your master [Protagoras] that he must admit that one man is wiser than another and that the wiser man is the measure, whereas an ignorant person like myself is not in any way bound to be a measure. (*Theatetus*, 179b)

Then Plato goes on to question whether even private sensations, "what the individual experiences at the moment" (*Theatetus,* 179c), are valid knowledge.[7] For Plato, as for the behaviorist, what is private is not knowledge and what is knowledge is not private. The reason for insistence on this is the same for Plato as for the behaviorist: that a person's mind (and soul) consists of actions of the person as a whole, not parts of the person. Actions of a whole person (like the actions of a state) cannot be intrinsically private.

In summary, Plato was the great exponent of what today we would call *context.* On a football team all the other players constitute the context of each player. Take a player out of the team and you have (a model of) an ideal form, a gap, that no particular player could perfectly fill. Take all of the particular players away and you have a wholly ideal team, a subworld of forms with no particulars at all.

When we behave in accordance with particulars (when we are guided by our pleasures) we are, like the prisoners in Plato's allegorical cave, interacting with shadows. When we behave in accordance with the *forms* (when we are guided by goodness rather than pleasure) we are, like the freed prisoner, interacting with reality.

For Plato, the abstract world of *forms* existed prior to (and is responsible for) the world of particulars (as three-dimensional reality is responsible for any particular two-dimensional projection). It was Plato's insistence on the ontological priority of the world of *forms* that was Aristotle's taking-off point. For Aristotle, to whom we turn next, abstractions retain the exalted status they had in Plato's philosophy but an abstraction cannot exist prior to a particular—there can be no team without players, no context without text.

Notes

1. This chapter is a revision of a previously published article (Rachlin, 1985). In it, I refer frequently to the three volumes of commentaries by Paul Friedlander (1958, 1964, 1969) not because I believe his views to be generally accepted but because his analysis is by far the most thorough, detailed, and comprehensive available in English.

2. However, Skinner (1974) disagrees that Plato's philosophy is susceptible to behavioristic interpretation. He has opposed Plato's "invention of the mind" to his own behavioristic viewpoint.

3. In Aristotelian terms, Plato sees goodness as a *formal cause* of reality. Aristotle distinguishes between formal causes in static systems and final causes as functions in dynamic systems, but Plato does not make this distinction. A function or an end for Plato is a higher level of abstraction (a formal cause) of the thing itself. Thus, for Plato even more than for Aristotle, the function of a thing stands "above" the thing. To put it another way, for Plato, reality is merely a projection of its function: goodness.

4. A similar diagram has been used in human decision theory (Machina, 1987) to represent three probabilities that sum to unity. In Machina's diagram,

however, the highest point is not in the middle but at one of the corners of the triangle. In economic terms, the circles are iso-utility contours and the triangle represents a constraint (a budget). Thus, economics, which, at an underlying level, is a theory of mutual benefit through interpersonal relations (Becker, 1976), would have been conceived by Socrates not as the "dismal" science but as the *veritable* science of love.

5. Near the beginning of *The Republic* (II, 361) the dialogue touches on the difference between a man who is just and a man who seems just. It is possible to consider this as a distinction between a man who is internally just and a man who is externally just. But Plato insists on viewing the distinction as one between overt action (being just) and public reputation (seeming just). Plato decides to look for the meaning of justice in the state before considering its meaning in individuals because "the city [is] larger than man." (II, 386e). It is also true that, while a man may easily keep much of his overt behavior from public view, a city may do so only with great difficulty. Thus, in a city, reputation and action more nearly coincide.

6. This "whole body" theory of learning is the same as the "personal level" theory rejected by Dennett (see p. 29). Whether a personal level theory is worth pursuing, or deserves the name "psychology" (which I take to mean, whether it is a theory of the mind), readers must judge for themselves.

7. But Socrates says in this passage that "it is harder to assail the truth of these [judgments based directly on sensation]." He goes on to admit: "Perhaps it is wrong to say 'harder'; maybe they are unassailable, and those who assert that they are transparently clear and are instances of knowledge may be in the right." Here, again, Socrates is being ironic. Yet he never does bring up the subject again in this dialogue. It is left to Aristotle to come to grips with the problem of the relation of sensation and knowledge.

4

Aristotle's Scientific Method

Imagine Aristotle in Athens. His former pupil, Alexander the Great, sends him crate upon crate of specimens: whole living animals, dead animals, parts of animals, dried plants, living plants, stones, vials of strange soil, etc. The crates arrive and Aristotle opens them and begins to sort them out. Imagine a vast combination warehouse, museum, and zoo with many bins, sub-bins, sub-sub-bins, and so forth. Aristotle develops a system for classifying specimens, putting them in bins, and labeling the bins. Later on, in addition to real animals, vegetables, and minerals, Aristotle begins to classify and store written descriptions of animals, vegetables, and minerals. Still later, he begins to classify and store not only actual things, not only descriptions of what things look, feel, smell, and sound like, but also descriptions of how they behave—that is, what they do.

The works of Aristotle are a key, a guide, a classification system, a set of bin labels for such (an imaginary) museum and library.[1] Aristotle's classifications (the bins) correspond in his system to the forms in Plato's system. J. H. Randall, Jr., a modern interpreter of Aristotle (1960, p. 51), quotes John Dewey as follows: "Classification and division are [for Aristotle] counterparts of the intrinsic order of nature."

Aristotle and Plato

On one hand, Aristotle is much easier to interpret than Plato. The works of Aristotle have come down to us in the form of straightforward monographs, organized by topic. There is no personal drama in Aristotle's basic works. Aristotle is much more sparing in his use of irony and myth (although more deliberately metaphorical) than is Plato. All of these factors seem to make Aristotle accessible. On the other hand, it is easier to get lost in detail in Aristotle and, because Aristotle was so specific in his observations, to infer apparent methodological errors from his factual disagreement with modern thought (his belief in spontaneous generation or the infinite speed of light, for instance). Aristotle does seem outmoded in a sense. His corpus constitutes so precise a description of the ancient Greek world that it seems no longer relevant; Plato's dialogues, on the other hand, by their very ambiguity, seem more applicable to current problems.

The point of this chapter is that, despite some real discrepancies between Aristotle's and Plato's philosophies, the two bodies of work can profitably be viewed, from the present, as a continuous and consistent whole. Randall (1960, p. 21) has this to say: "There is not an Aristotelian document in which Platonic insights are not in some form deeply imbedded." Certainly Kantor (1963, p. 161) saw things this way; he places Aristotle at the very apex of "naturalism" and Plato only a notch below. The contemporary neo-Aristotelian philosopher Alan Donagan (1987) bases his discussion of Aristotle's teleology on Plato's earlier holistic conception of human behavior.

The founder of contemporary hermeneutic philosophy, Hans-Georg Gadamer (1986), argues that Aristotle's idea of the Good is perfectly consistent with Plato's. Gadamer claims that it is a mistake to see the work of either Plato or Aristotle as a public development or a learning experience of the author. Rather, what seem to be earlier ideas rejected and supplanted by later ones are really basic principles (*archae*) upon which the further development rests (as a ship rests on its keel, to use an Aristotelian metaphor). Plato's concept of self-control (of behaving well), illustrated in Figure 3.1, would be viewed by Gadamer as underlying the wider view of Figure 3.2, just as arithmetic might be viewed as underlying calculus. Correspondingly, Aristotle, who was a Platonist in his early years, never abandoned Plato's idea of the Good but rather developed that idea significantly. "The task," according to Gadamer (p. 21), "is to get back to the common ground upon which both Plato and Aristotle base their talk of the *eidos* [the eidetic entities, or forms]." Gadamer's ultimate goal is to find a role for ancient philosophy in modern life. He says (p. 6):

> I ask that the reader take what follows as an attempt to read the classic Greek thinkers the other way round as it were—that is, not from the perspective of the assumed superiority of modernity, which believes itself beyond the ancient philosophers because it possesses an infinitely refined logic, but instead with the conviction that philosophy is a human experience that remains the same and that characterizes the human being as such, and that there is no progress in it, but only participation. That these things still hold, even for a civilization like ours that is molded by science, sounds hard to believe, but to me it seems true nonetheless.

We will return to the issue of the relation of philosophy, psychology, and modern science in Chapter 7.

As Gadamer emphasizes, the similarities between Plato's and Aristotle's philosophies are extensive and fundamental while the differences are narrow and secondary. Nevertheless, having seen Plato from a behavioral viewpoint, we may more easily see Aristotle from that same viewpoint by stressing their differences. This is nothing but the application of the dialectical method, endorsed by both philosophers, to their works themselves.

There is an unresolved contradiction at the core of Plato's philosophy. For Plato, as we have seen, nothing surpasses living a good life. In an imperfect (an actual) society even the best of people are faced with difficult choices: they must calculate, they must provide detailed accounts of their opinions, they must be philosophers. A perfect society, however, would be explicitly designed to encourage all people to live a good life. In such a society it would be easier to make choices. It would not be necessary to calculate, to provide detailed accounts of opinions, because the society itself (the laws of the state plus the leadership of a philosophical elite) would perform these functions. In a perfect society a person could be good without being wise.

Plato's order of priorities implies that true opinion (identified with good behavior) by itself is better than an account (a verbal rationalization of opinion) by itself (although both together are better still; both together constitute knowledge). With true opinion alone, Plato says, people are like "blind men who go the right way" (*Republic* VI, 506c). However, if society were to become perfect, Plato implies, most people, as individuals, could lead good lives without an (individual) account. The account would, in a sense, be disembodied from individuals and re-embodied in society. This way of looking at the relationship between good behavior (true opinion) and reason (the account) must have bothered many Greeks. It certainly bothered Aristotle. The *Nichomachean Ethics (NE)* is an attempt to provide an alternative view. (Another attempt, by St. Augustine, will be discussed in Chapter 6.)[2]

For Plato, the Good is absolute. Reason is tied to goodness only in an imperfect society. To the degree that society becomes perfect, the tie that binds reason to goodness (the need to make difficult choices) loosens in the individual and, while goodness remains in the soul of the individual, reason becomes identified with the soul of society itself. Utopian societies from *The Republic* through Aldous Huxley's (1946) *Brave New World,* to Skinner's (1948) *Walden Two* have been criticized on this ground. (In many imaginary ideal societies an elite group or a single individual "manager" continues to make decisions regarding policies and laws. However, as Skinner has pointed out, if the society is functioning well, a manager's behavior, although perhaps more complex, is no less determined by the *society's* rules than is the behavior of other citizens.) This was a problem for ancient Greece as well as for contemporary Western societies because it apparently took responsibility away from individuals. Because the laws of the state in Greek society as well as our own are generally concerned with punishment and (occasionally) reward of individuals based on each person's responsibility for his or her actions, Plato's ideal state seems inconsistent with Western law. As both Plato and Aristotle recognized, philosophy inconsistent with law is inconsistent with the fundamental ideals (the *archae*) of society.

Aristotle's resolution to Plato's contradiction between collective and individual goodness is to shift the focus of fundamental philosophical interest from ethics to epistemology—from goodness to knowledge. Aristotle says of Socrates that he "was busying himself about ethical matters and neglecting the world of nature as a whole" (*Metaphysics* I, chap. 5, 987b, 1). Randall says (1960, p. 3), "For Aristotle the highest power a man can exercise over the world is to understand it—to do, because he sees why it must be done, what others do because they cannot help themselves." The *Metaphysics* (*M*), Aristotle's most abstract work—the one in which the most fundamental principles are set forth—begins with: "All men by nature desire to know" (*M,* A1, 980a, 1).

As things are, knowledge is as much tied to goodness for Aristotle as it is for Plato. Less-than-perfect human beings in a less-than-perfect society cannot be intelligent without also being good. Aristotle says, "No one deliberates [exercises reason] about things that are invariable, nor about things that it is impossible for him to do" (*NE,* VI, 5, 1140a, 32). However, as Randall (1960, p. 79) says, "The *Ethics* ends by making clear Aristotle's own preference for the life of *theoria,* that sheer knowing in which man transcends the limits of human nature and becomes like the gods." To summarize, both Plato and Aristotle see the best life as a unity of goodness and reason. For Plato, a person may behave well without an individual account only in a perfect society. For Aristotle, pure reason (reason without reference to good behavior) may be exercised only by a perfect person. As long as we stick to things as they are (imperfect societies and people), Plato and Aristotle agree: living well is inseparable from reasoning well.

The behavioral view of Plato taken in Chapter 3 is, I emphasize again, distinctly a minority view. A behavioral view of Aristotle, however, is not so rare and may not seem so strange to the reader. If it is not exactly a majority view among Aristotelian scholars, at least some influential Aristotelian philosophers hold it or have held it. According to Randall (1960, p. 66): "For Aristotle life or *psyche* is the behavior of the organism as a whole in its environment. Aristotle is thus a thoroughgoing behaviorist." One reason that Aristotle is a behaviorist is because, according to him, reasoning is not sitting in a chair with your eyes closed, but is behaving—arguing. Aristotle agrees with Plato that reason cannot be set off against overt behavior (as Danto implies in the quotation on p. 44) but is itself a form of overt behavior. The difference is that, building on Plato's demonstration of the truth of this fact, Aristotle is concerned with discovering and with marking off from everything else, exactly what sort of behavior good reasoning is.

The Forms

For Plato, forms are functions (like ecological niches), while for Aristotle forms are classificatory divisions. The distinction is slight because a classification for Aristotle is almost the same thing as a function for Plato. The difference is really one of emphasis and rests on three related points. First, an efficient system of classification would have no empty bins. Therefore, for Aristotle, there are no forms without objects; if all the chairs in the world were to disappear there would be no category of a chair. Second, the forms of things, while certainly longer-lasting than individual things, are not eternal. Aristotle says in *Posterior Analytics* (*PA*, I, chap. 22, 83a, 33): "The Forms [Plato's eternal forms] we can dispense with, for they are mere sound without sense [language without meaning]." Third, while Plato ties the concept of reality, of actual existence, to the concept of permanence, and concludes that only forms are real because forms are eternal while particular objects come and go, Aristotle recognizes various degrees of reality.

Aristotle is careful, as a categorizer must be, to keep the levels of categorization distinct. He says (*PA*, I, chap. 24, 85a, 30): "The universal has not a separate being over against groups of singulars." Here he is warning against what Ryle (1949) was to call a category mistake. A boy who was promised a dog for his birthday, and was given a poodle, and who then said, "Thanks, but where is my dog?" would (discounting possible irony) be making a category mistake. He would be putting the universal over against a particular as if they constituted separate beings. Another reason why Aristotle is a thoroughgoing behaviorist is that for Aristotle the relation of mind to behavior is the same as the relationship of universal to particular. If the relation of mind to behavior is that of universal to particular, observation of a person's behavior would be both necessary and sufficient for observation of that person's mind. (I shall discuss this contention in some detail later.) While, for Aristotle, mind and behavior both are real in a sense, they are each real in different senses. Mind is real in the sense that dogs are real. Behavior is real in the sense that poodles are real.

It is important to note that Plato does not make a category mistake, nor does Aristotle accuse him of making one. Plato does not put particulars on the same level as universals; instead he denies the essential reality of particulars. Aristotle agrees with Plato that the reality of universals is higher in some sense than that of particulars and—for Plato's very reason—that universals are less transitory. In arguing for the superiority of universal definitions (like the definition of a triangle) over particular ones (like the definition of a poodle), Aristotle says (*PA*, I, chap. 24, 85b, 15): "The universal will possess being not less but more than some of the particulars, inasmuch as it is universals which comprise the imperishable, particulars that tend to perish." The difference between Aristotle and Plato then boils down to the question of whether particulars have any "being" at all. For Plato, they do not (and the belief that they do is what prevents people from living well). For Aristotle, particulars do have a kind of being because only through particulars can universals be grasped.

In strictly behavioral terms Plato and Aristotle both view abstract knowledge as a form of behavior, as a pattern of behavior matching the higher levels of reality (forms or universals). For Aristotle, these abstract patterns require particulars (as a jigsaw puzzle requires pieces), while for Plato they exist independent of particulars (as the pattern of a jigsaw puzzle exists prior to its being cut). To take a mundane example, consider knowing how to ride a bicycle. In a sense, most normal people can ride a bicycle even if they have not yet learned, even if no bicycles existed. This is the sense in which we all "know" everything before being born. In *De Anima (DA)*, Aristotle agrees with Plato that we all know everything this way, "the sense in which we might say of a boy that he may become a general" (*DA*, II, chap. 5, 417b, 33). But Aristotle points out that there are degrees between this sort of potentiality and actuality. There is a difference between the potential of a boy to become a general and the potential of a man to become a general. Similarly, when a person is actually a general, there are differences between being a general while asleep and while awake and, if awake, while entertaining friends and while calling out orders to his troops. Aristotle finds these sorts of distinctions essential to classify different sorts of behavior. But if these sorts of distinctions are to be made, he argues, attention must be paid to particulars.

To take another example, on a more exalted level, consider the difference previously discussed between societal and individual rationality. Aristotle agrees with Plato that a person cannot be rational without behaving rationally. However, for Aristotle, rationality is the pattern of behavior that all rational people have in common (which Aristotle takes pains to define as precisely as possible), while for Plato rationality is a pattern in the universe (like an ecological niche) that would exist even if there were no rational people—even if there were no people. With regard to immediate behavior, this is no difference at all. A group of people all behaving rationally in Aristotle's sense (by dominance of their individual autonomous rational souls) would be doing exactly the same thing as a group of people all behaving rationally in Plato's sense (by slavish obedience to the *form* of reason). Nevertheless, Aristotle's way of looking at such behavior not only seems more in accord with the modern Western viewpoint than Plato's does, it also seems to make real differences to the rest of his philosophy. Whereas Plato was concerned with the pattern of goodness in the world, Aristotle was concerned with describing rationality as it appears in actual human behavior.

Aristotle's earliest works define terms and lay the groundwork for the detailed categorizations to come. In *Prior Analytics,* Aristotle codifies the rules of deductive logic (describes a tool). In *Posterior Analytics,* he shows how logic may be applied to the world (shows how the tool should be used). In other words, *Posterior Analytics* outlines Aristotle's conception of scientific knowledge.[3]

Logic

Aristotle's deductive logic is a formalization of Plato's dialectic. Recall that the dialectic is a method of argument. Two people agree about certain basic statements but disagree about certain contradictory particular statements. Dialectics is the art of demonstrating that one of the particular statements follows from the agreed-upon basic statement and one does not. One of the original disputants is shown thereby to have contradicted himself.

Posterior Analytics (*PA*, I, chap. 1, 71a, 1) begins:

> All instruction given or received by way of argument proceeds from pre-existent knowledge. This becomes evident upon a survey of all the species of such instruction. The mathematical sciences and all other speculative disciplines are acquired in this way, and so are the two forms of dialectical reasoning, syllogistic [deductive] and inductive; for each of these latter makes use of old knowledge to impart new, the syllogism assuming an audience that accepts its premises, induction exhibiting the universal as implicit in the clearly known particular.

The difference between Aristotelian science and Platonic dialectical argument is that in Aristotelian science the *archae* (basic principles, "old knowledge") are supposed to be really true while in ordinary Platonic dialectics (in argument) the basic principles are just generally accepted by the people arguing (Randall, 1960, pp. 40–41). But what does it mean for something to be really true for Aristotle? Again we come to Aristotle's emphasis on individuality. For him real truth is to be found in individual perception—in the exercise of a human's highest faculty *(nous),* the ability to see the general in many instances of the particular: "When the observation of instances is often repeated, the universal that is there becomes plain" [Randall's (1960, p. 43), translation of *PA,* II, chap. 19: 100a 5, 6]. Thus, the most fundamental truths for Aristotle are products of a process in which individuals interact with the world over a period of time. The universal arises out of many particular events "like a rout in battle stopped by first one man making a stand and then another, until the original formation has been restored" (*PA,* II, chap. 19, 100a, 11).

Ideally, science begins with such truths and derives further particulars from them. As Aristotle applies it, science begins with observation of particulars, generalizes from them, and only then applies deductive logic. Returning to previous imagery, it is as if Aristotle, when the first specimens began to arrive, created a set of temporary holding bins until sufficient numbers provided perceptible general characteristics (until "a stand" could be made). Then more or less permanent bins were constructed and labeled, and future specimens properly classified.

Let us assume for the moment that the initial tentative process is complete, that the permanent bins have been constructed and labeled, and that a new specimen arrives. Where does it go?

Consider the following syllogism:

A. All poodles are dogs.
B. This thing is a poodle.
Therefore: C. This thing is a dog.

A new specimen that can be identified as a poodle goes in the dog bin (which is a subcategory of the animal bin). Figure 4.1 puts the syllogism in diagrammatic form.

Statement A, "All poodles are dogs," is the major premise of the syllogism. It is a statement of a universal fact. The conclusion C, "This thing is a dog," is a statement of a particular fact. Since, as we have said, Aristotle views the relation of mind to behavior as that of universal to particular, it is worth examining his view of the relationship of statement A to statement C. First, reasoning is viewed as a kind of learning ("instruction given or received by way of argument") where what you already know is a rule (you've already fixed and labeled the bins) but what you do not know is that this particular case is an instance of that rule. The syllogism tells you whether it is. Plato had said that you cannot learn without in some sense already knowing what you are about to learn (recall the slave boy in Plato's *Meno* who "already knew" what he was taught). Aristotle agrees that you cannot be taught anything without already knowing something. You already know the general rule. And, in a sense (not the same sense in which you know the rule), you know the particular. What you do not know (when you start) is whether this particular is a case of this rule. For example (*PA*, I, chap. 24, 86a, 24), suppose you *actually* know that the sum of the angles of all triangles is 180° (in the sense that you have demonstrated that fact). Then you *potentially* know that the sum of the angles of an isosceles triangle is 180° even though you may not know what an isosceles triangle is. Once you learn that an isosceles triangle falls under

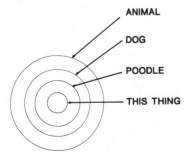

ANIMAL

DOG

POODLE

THIS THING

Figure 4.1 A set of categorical inclusions.

the general class of triangles, your potential knowledge becomes actual, using the syllogism:

D. The sum of the angles of all triangles is 180°.

E. An isosceles (triangle) is a triangle.

Therefore: F. The sum of the angles of an isosceles triangle is 180°.

Because you do already know the universal, learning by induction (perception of universals through particulars) must have preceded learning by deduction (demonstration). Therefore, "the premises of demonstrated knowledge [the universal] must be true, primary, immediate, better known than and prior to the conclusion [the particular]" (*PA*, I, chap. 2, 71b, 20).

Since induction must have preceded deduction, all knowledge rests on the inductive process—the direct perception of universals that Aristotle says must come through particulars. What could it mean to see universals through particulars themselves? Perhaps the closest modern concept to this conception is that of perceptual *transparency*. When you look at a scene through a telescope, for instance, the telescope critically affects what you see but you do not see the telescope, you see the scene. In this sense the telescope is transparent. As another example, consider what you see when you watch a movie. The colored lights and shadows are necessary for you to see anything, but when you leave the theater, you do not discuss lights and shadows but the behavior of characters. The lights and shadows are in this sense transparent; you see "through" them to actual characters. Going a step further, you might even see through the behavior of the characters in a movie to their mental states. You might say, for instance, about a character in a movie, "She had no right to be as happy as she was," and yet not be able to say, even upon subsequent analysis, exactly what she did. Still less would you be able to discuss her emotions in terms of light and shadow.

For Aristotle, the grasping of universals through particulars without awareness of particulars is like seeing a character's mental state (through behavior) without necessarily perceiving the behavior itself. Although particulars might not be perceived as such (although a telescope or the lights and shadows on a movie screen might not be perceived as such), it is nevertheless impossible, according to Aristotle, to grasp universals without them. For Plato, on the other hand, the forms *do* exist without particulars just as the characters and their emotions might exist in a script before the movie is made. An actual movie with actual characters might then be conceived as an approximation to some ideal performance embodied in the script. Going back a step, we could conceive of these ideals in the author's experience before existing in the script, and in the world before experience.[4] What you see in a movie might then be conceived—the way Plato would—as only a pale and inadequate "projection" of this ideal.

Cause and Necessity

Immediately after distinguishing between the primacy of universals relative to particulars, Aristotle says that the conclusions of demonstrated knowledge (the particulars) "are related to the premises [the universal] as effect is to cause" (*PA*, I, chap. 2, 71b, 22). In other words, the premises are in some sense the cause of the conclusions. If Aristotle means, by "cause," as we generally do, an efficient cause, then Aristotle can be said to see the mind as causing behavior in that way (as an efficient cause). But, like Plato, Aristotle was writing before Renaissance science had legitimized only efficient causes. Aristotle's broadest conception of "cause," like Plato's, is logical. For Aristotle, an ordinary cause is whatever can serve as a middle term in a syllogism—poodles and triangles in the cases above. The middle term is what usually follows "because" in a logical argument (See Chapter 2). Thus, this thing is a dog *because* this thing is a poodle (and all poodles are dogs). Aristotle says, "In all our inquiries we are asking either whether there is a 'middle' or what the 'middle' is: for the middle here is precisely the cause, and it is the cause that we seek in all our inquiries" (*PA*, II, chap. 2, 90a, 5). A dog is potentially a cause because it may be a middle in a syllogism with the major premise: "A'. All dogs are mammals," or "A". All dogs are animals."

For Aristotle, a cause can be more or less scientific. The most scientific causes are the most abstract; for instance, the geometrical axioms. The following syllogism:

> A". All dogs are animals.
> B". This thing is a dog.
> Therefore: C". This thing is an animal.

is more scientific than the one discussed on p. 73 because the middle term of this one (dog) is more abstract (further from the center in Figure 4.1) than is the middle term of the other (poodle). A middle term (a cause) of one syllogism may be a conclusion (an effect) or a major premise of another. (A category may be a unit of a higher category.) Thus, the relation of the universal to the particular is relative. Within a given science, universals, like numbers in arithmetic, which are universals with respect to units, may be treated as particulars—as particular even numbers, for instance. The relativity of the universal-particular relationship serves in turn for Aristotle as an argument against the existence of universals separate from particulars.

As larger and larger circles are drawn, the fundamental principles (the *archae*) of a science are eventually reached. The truth of fundamental principles (the axioms of geometry, for instance) cannot be demonstrated deductively because there are no higher categories (larger circles) into which they fit. These fundamental truths are not properly part of the particular science itself (as a carpenter's tools would not be made by a carpenter). Aristotle says (*PA*, I, chap. 11, 77b, 5), "Of the basic truths [of geometry] the geometer, as such, is not bound to give any account."

For Aristotle, the fundamental principles of a science are "true, primary, immediate, better known than and prior to" individual sensory events. Again, this is opposite to the way we think of these issues today. We have taken our paradigm cases of truth, primacy, immediacy, certainty, and priority from the British associationist philosophers. For them individual sense impressions, "raw feels" (pains, for instance), epitomize these categories. For Aristotle, mental events are truer, more fundamental, more immediate, more certain, and prior to physical events, not because they are closer to individuals (smaller circles in Figure 4.1), still less because they exist only within individuals, but on the contrary, because they are further away (larger circles), hence, more universal. Aristotle makes this point explicitly (*PA*, I, chap. 2, 72a, 1–10):

> Now "prior" and "better known" are ambiguous terms, for there is a difference between what is prior and better known in the order of being and what is prior and better known to man . . . the most universal causes are furthest from sense and particular causes are nearest to sense and they are thus exactly opposed to one another.

Perhaps Gustave Flaubert (1856–1857/1950, p. 49) was referring to this same distinction when he described Madame (Emma) Bovary's character as follows: As an adolescent, "she rejected as useless whatever did not minister to her heart's immediate fulfillment—being of a sentimental rather than an artistic temperament, in search of emotions, not of scenery." Flaubert is here holding up the narrow (emotional) context of Emma's choices as the key flaw in her character—the trait that is going to get her into trouble later in life. To anticipate Chapter 5, Aristotle will use this logical distinction, between universal and particular, as a guide to ethical behavior.

The most scientific causes are the most universal. These are *formal* causes. In what sense are formal causes superior to other kinds? The more universal the cause utilized by a given science, the more permanent are the objects of that science; hence, the truer the principles. Despite Aristotle's continuous focus on the individual, he agrees with Plato that reason properly aims at universal truth. As Gadamer (1986) points out, Aristotle does not object to Plato's focus on the universal. What Aristotle does object to is the lack of connection in Plato's philosophy between the universal and the particular.

Aristotle, in discussing other philosophers' views of causation in the first book of *Metaphysics,* argues that *formal* causes are superior to *material* causes (the notion that something is what it is because of its material composition—the elements, air, earth, fire, and water) because formal causes stand to material causes as universal to particular. If there were no change in the world, these two causes would suffice. However, there is change in the world and neither material nor formal causes can explain it. (Aristotle accuses Plato of using only these two causes, hence failing to explain physical change.)

When Aristotle applies logic to physical change over time (*PA*, II, chap. 11; *Physics* II, chaps. 3, and 7), he discusses two other causes, *final* causes (the aim of the change—"for the sake of what") and *efficient* causes ("what originated a motion"). Final causes stand to efficient causes—as formal causes stand to material causes—as universal stands to particular. Ultimately, the reason that all four types are causes is that all four may serve as the middle term of a syllogism. (Efficient and final causes are in the middle temporally as well as logically.) Because the syllogism is the "account" that must go along with true opinion to form knowledge, "it is the business of the physicist to know them all" (*Physics* II, chap. 7, 198a, 23). Since, for Aristotle, psychology is a branch of physics, he would say that it is the business of the psychologist to know them all. For Aristotle, discovery of the efficient causes (physiological or cognitive) of animal (including human) behavior would not preclude search for the final causes (reinforcers, in the widest sense of that term, discussed in Chapter 2) as the mere hope of such discovery threatens to do today.

In the world of ancient Greece physical laws (except those governing movement of celestial bodies) had none of the cachet of *necessity* that they bore in modern times until very recently. According to physics, from the Renaissance through the nineteenth century, if you push one side of the seesaw down, the other side *necessarily* goes up because the seesaw obeys various physical laws. For Aristotle (as well as Plato), on the other hand, *necessity* (below the level of celestial bodies) is reserved for logic. A seesaw would fall into the class of objects which if you push one side of them down, the other goes up. If the other side were not to go up, you would not have been pushing a seesaw. Therefore, when you push one side of a seesaw down, the other side *necessarily* goes up because logic says so, not physics. (Logic, of course, would be pointless without underlying physical lawfulness. Were it not for common physical characteristics among its members, the category "seesaw" would have no application. Nevertheless, the *necessity* underlying the transition from cause to effect is conceived by both Plato and Aristotle as fundamentally logical.)

Form and Substance

Imagine again Aristotle pursuing the classification of descriptions in the library and in the museum (which has eventually become the world itself). Although what is in the library describes what is in the museum, it is easy to conceive that different ways could develop of speaking about the contents of the library (speaking about language) and speaking about the museum (speaking about the world, excluding language).

Randall (1960), in discussing Aristotle's *Metaphysics,* distinguishes between a formal analysis and a corresponding functional analysis. He says (p. 59) that "things can be understood in the way they can be talked about, and they can also be understood as taking place in a dynamic process of change." Before proceeding to our main concern—Aristotle's functional accounts of human behavior in *De Anima* and *Nicomachean Ethics*—it is necessary, following Randall, to distinguish between these two types of analysis.

In the formal analysis (the organization of the library) " 'to be' anything means 'to be something that can be stated in discourse.' " In the functional analysis (the organization of the world) " 'to be anything . . . means 'to be something that comes into being and passes away,' something that is subject to change, that persists throughout a determinate change" (Randall, p. 111). Logical demonstration (the syllogism) is the vehicle of both forms of analysis. Furthermore, functional analysis is itself discourse and therefore may be formally analyzed. (That is, functional analyses go in the library where they must then be classified according to the system there: formal analysis.)

In formal analysis, to be something that can be stated in discourse means to be *definable;* in a definition, a *subject* is said to have certain *attributes* that distinguish it from all other subjects. In functional analysis, to be something that persists through a determinate change means to be a *substance;* in a substance, *matter* is said to have a certain *form.* Figure 4.2 illustrates the logical, formal, and functional relationships involved.

Figure 4.2 is drawn differently from Figure 4.1 in an attempt to indicate that in this figure there is no real space between the circles. In a definition the essential attributes, which Aristotle (*PA,* I, chap. 4, 73b, 26) also calls "commensurate universals," are intended to so delimit the subject that no other subject could fit in. There are dogs other than poodles and animals other than dogs. Thus, Figure 4.1 is drawn with spaces between the circles. But, imagining that an apple were the only round red fruit, the intersection of the attributes round and red "in" the subject (fruit) would define that fruit as an apple.[5]

In Figure 4.2 the boundaries of Figure 4.1 have collapsed into one boundary. Figure 4.2 is a sort of limit of Figure 4.1 as the middle shrinks in extent. However, the order of the three elements (particular, middle, and universal) remains as it was. Despite the single boundary, in functional analysis, the form is in some sense closer to the substance of a thing than the matter is. The form is what remains the same through material change. For instance, a circle is a substance if matter (say, ink on paper) takes on certain form (say, $x^2 + y^2 = r^2$). Both matter and form are necessary. The

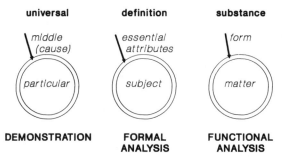

Figure 4.2 Corresponding concepts in logic (demonstration), formal analysis, and functional analysis. The inner circle is assumed equal in extent to, although conceptually within, the outer circle.

equation alone is not a circle, it is only the form of one. Yet $x^2 + y^2 = r^2$ is more like what a circle really is than is ink on paper. Matter, considered above, is *potentially* various substances and becomes *actually* one of them only when it takes on that particular form. A form considered alone can, however, only be one kind of substance. It is true that just as ink on paper can take on an infinity of forms, the equation $x^2 + y^2 = r^2$ can become matter in an infinity of ways (ink on paper, a cardboard disk, the boundary between the sun and the sky, etc.) But while the substance (the essence) of a given chunk of matter may be anything that the matter is potentially capable of being, the substance of a given form can only be one essential thing.

The (atemporal) formal and material causes belong properly to formal analysis, while the final and efficient causes, both of which impose temporal order on events, belong properly to functional analysis.[6] Just as, in formal analysis, the formal cause is superior to the material ("prior and better known in the order of being"), so too, in a functional analysis, the final cause is superior to the efficient. As I said previously, a cause for Aristotle is what follows the word "because" in an explanation. A formal or final cause is an answer to the question, "Why?" If a person keeps asking "Why?" to whatever answer you give them (as a child often does), the point where you run out of answers (and perhaps tell the child to get lost) is when you have reached the defining attributes of a formal analysis or the end of a process in a functional analysis. When you answer in this way (in terms of formal or final causes), you progress from the separated circles of Figure 4.1 to the unified ones of Figure 4.2. Going this way, toward the universal, demonstrations eventually come to a "simple and determinate" end. However, going the other way, toward the particular (if you answer in terms of material or efficient causes), "demonstration . . . sinks into an indeterminate manifold" (*PA*, I, chap. 24, 86a, 5).

To return briefly to present-day behaviorism, the distinction between efficient and final causes in Aristotle's functional analysis corresponds, in a limited way, to Skinner's (1938) distinction between respondents and operants. A respondent for Skinner is an act defined by its antecedents (as the characteristics of pupillary contraction are defined as functions of light in the eye), while an operant is an act defined by its consequences (as the characteristics of a rat's lever presses are defined as functions of reward and punishment). Like Aristotle, Skinner saw these relationships in behavioral rather than physiological terms. Skinner defines a single respondent (or a single operant) as a single correlative relationship between a prior (or subsequent) environmental event and an act of the organism (which may or may not be internally mediated by a single physiological pathway. For Skinner as for Aristotle, physiological relationships are accidental to behavior.)

The critical difference between Aristotle and Skinner is that Skinner (perhaps in fear of accusations of teleological thinking) insisted that a respondent (or an operant) is a particular behavioral event immediately preceded (or immediately followed) by a particular environmental event. An individual reinforcer, an individual food delivery to a hungry rat, was not conceived by Skinner as a final cause; its effects are future effects—on the subsequent rate of emission of the operant that the reinforcer has followed. It is true that there exists a gap in time between a reinforcer and the consequent rate increase but nevertheless the reinforcer comes first; Skinner therefore could conceive of an individual reinforcer as an *efficient* cause of an increased rate or probability of response.

Aristotle, on the other hand, had a much broader conception of behavioral and environmental events. Randall (1960, p. 67) says, "In some respects, present-day gestalt psychologists are closer to Aristotle than any other modern school." For Aristotle, as for the gestalt psychologists, reality lies in the universal, which may be directly grasped by the human mind. But what does it mean for something to be grasped by the human mind? For a gestalt psychologist, being grasped by the mind is a phenomenological event, a direct but wholly private perception. A behaviorist has to say that being grasped by the mind is *nothing but* being exhibited in overt behavior. I shall discuss Aristotle's opinion on this subject in Chapter 5.

Change

A functional analysis is, as Randall says, an analysis of "what persists through a determinate change." Before discussing change as it applies to humans, we need to understand Aristotle's conception of change itself (as presented in *Physics* V and VI). It is, like ours, very broad. However, his concept of motion is also very broad (unlike ours). Our everyday concept of motion, inherited from Renaissance physics, includes only locomotion (local motion): movement from place to place. For Aristotle, "there are three kinds of motion: qualitative, quantitative, and local" (*Physics* V, chap. 1, 225b, 8). Motion is a kind of change that includes change in quality (from black to white, from health to sickness) and changes in quantity (growth of a person or a tree, for instance) as well as local movement (of objects through space). In addition to this wide definition of motion, there are two types of change that are not motions: coming to be (change from a nonsubject to a subject in formal analysis) and passing away (change from a subject to a nonsubject in formal analysis).

Thomas Kuhn (1987) first became aware of the nature of scientific evolution when, as a graduate student in physics, he was preparing a course on science to teach to undergraduate nonscience majors. He felt that modern physics might be clarified by tracing its origins in ancient science, saying (p. 9):

> The question I hoped to answer was how much mechanics Aristotle had known, how much he had left for people like Galileo and Newton to discover. Given that formulation, I rapidly discovered that Aristotle had known almost no mechanics at all. Everything was left for his successors, mostly those of the sixteenth and seventeenth centuries. That conclusion was standard, and it

might in principle have been right. But I found it bothersome because, as I was reading him, Aristotle appeared not only ignorant of mechanics, but a dreadfully bad physical scientist as well. About motion, in particular, his writing seemed to me full of egregious errors, both of logic and of observation.

But then, Kuhn thought, how could such an acute observer of biological phenomena and such a well-respected thinker have been so wrong? Kuhn goes on:

> Feeling that way, I continued to puzzle over the text, and my suspicions ultimately proved well founded. I was sitting at my desk with the text of Aristotle's *Physics* open in front of me and with a four-colored pencil in my hand. Looking up, I gazed out of the window of my room—the visual image is one I still retain. Suddenly the fragments in my head sorted themselves out in a new way, and fell into place together. My jaw dropped, for all at once Aristotle seemed a very good physicist indeed, but of a sort I'd never dreamed possible. Now I could understand why he said what he'd said, and what his authority had been. Statements that had seemed egregious mistakes, now seemed at worst near misses within a powerful and generally successful tradition. That sort of experience—the pieces suddenly sorting themselves out and coming together in a new way—is the first general characteristic of revolutionary change. . . . Though scientific revolutions leave much piecemeal mopping up to do, the central change cannot be experienced piecemeal, one step at a time. Instead, it involves some relatively sudden and unstructured transformation in which some part of the flux of experience sorts itself out differently and displays patterns that were not visible before.

What was it that Kuhn discovered? Essentially it was Aristotle's functional analysis—the application of logic (of syllogistic demonstration) to change. It is in his analysis of change, Aristotle claims, where he goes beyond Plato.

For Aristotle, no motion could conceivably occur without something that moves (matter) and a pattern of movement (form). When the matter has completed the pattern, the process of movement is complete. The process may be subdivided into stages as building a house may be divided into building a foundation, building walls, building a roof, and so forth. Each stage may then be considered as an individual process subsidiary to the whole. Each stage consists of matter taking on a new form. Aristotle says (*Physics* VI, chap. 5, 235b, 6): "Since everything that changes changes from something to something, that which has changed must at the moment when it has first changed be in that to which it has changed." In other words, as a process moves from stage to stage, the earlier stages become part of the later ones; the process of building a foundation becomes part of building a house. Another example: an orchestra might be playing an opera overture as an individual piece in a concert or as part of the opera itself. Until the overture has been played it cannot even be said that the orchestra is playing an overture (it might break into "Take Me Out to the Ballgame" at any moment). Suppose the orchestra finishes the overture as written. Then it might just have played the overture (as part of a concert) or be in the midst of playing the whole opera. When the curtain rises after the overture and you see the set and characters of the first act, you may say (provisionally) that the orchestra is playing the opera. Only after the opera is complete can you actually say what happened: the orchestra has played an opera and the overture was part of it.

Words (logic) cannot fit exactly onto incomplete motions. If motions had no apparent end, they would not be classifiable. This issue is important for modern philosophy and psychology, which speaks of "intentional" acts. The concept comes from Brentano (1874 / 1973) who got it from Aristotle. But for Aristotle an intention would not be an internal state of the actor. It would be a provisional classification of a movement. Thus, for Aristotle, a rock as well as a person might have an intention. The rock may be part of an avalanche that will eventually destroy a town, or it may be just rolling down the mountain by itself (heading toward the center of the earth, its natural place). Which category it falls into (no pun intended) depends on its overt context—depends on what else happens—not on occult events within the rock.

For Aristotle, a man walking to the store to buy a loaf of bread, a rock falling down a mountain about to start an avalanche, an octagon with only seven sides drawn in, all have the same quality of incompleteness (or "intentionality"). In all three cases the matter in question is not identical with the form in question. The man's errand, the rock's falling, the octagon are not actually what they are said to be. All three may be classified only provisionally. The man may forget his errand; the rock may fall harmlessly to the bottom; the octagon may twist off into a spiral.

For Aristotle, any complete movement is defined by only two terms: the thing moved and the form of the movement. If the movement has not yet taken place, the matter may be said to be potentially of a certain form (as a black log is potentially white). But *movement is always defined without explicitly specifying a beginning*. In the case of qualitative movement, this poses no problem for us because it is there that our language conforms to Aristotle's. Like Aristotle, we view qualitative changes as processes with determinate ends. The beginning (or at least its direction) is implied by the end. If a log becomes white, it must have been darker. If a man becomes sick he must have been healthier. If a girl learns the Pythagorean theorem, she must have been ignorant of that theorem beforehand.

In the case of locomotion, however, the lack of a beginning seems strange. We say something moves from point A to point B. Aristotle says just that it moves to point B. If something moves to point B, it could have started from anywhere. In what sense, then, is the movement defined? The answer is that if locomotion is viewed as a natural process (as Aristotle views it) then it could not have started from *anywhere*. If the rock is heading toward its natural place, it must have started further from its natural place than it is now. The temporal origins of a given motion are approached by Aristotle not by going back to its efficient cause but by dividing it into natural subcategories and searching for the earliest final cause. Thus, building a foundation is the first end in building a house. But building a foundation may in turn be divided into stages, each defined by its end: surveying the ground, digging a hole, constructing wooden forms, filling them with rocks, and so forth. Surveying might be subdivided in turn. This process, however, could not go on indefinitely. Eventually, the description of the process would be complete.

The difference between locomotion and qualitative change (for Aristotle) is that qualitative change is, in a sense, reversible. That is, qualitative change could end in darkness as well as whiteness, ignorance as well as knowledge, sickness as well as health. Neither member of these pairs of opposites is more natural than the other. But in locomotion an object, unless blocked (unless part of some other definable process), always heads for its natural place (fire and air away from, liquids and solids toward, the center of the earth).

This perceived directionality is what seems so artificial to us. It seems as though Aristotle is taking a perfectly good explanation of qualitative change and imposing it upon locomotion (movement of objects in space) by artificially endowing locomotion with aims and motives. But Aristotle's object in endowing locomotion with ends (or, as he might have said, perceiving ends in locomotion) was to explain all kinds of change with the same set of principles. According to Kuhn (1987, p. 10): "It is precisely seeing motion as change-of-quality that permits its assimilation to all other sorts of changes—acorn to oak or sickness to health, for examples."[7]

Locomotion, like all other change, corresponds for Aristotle to a relationship between the universal and the particular. In moving, a particular thing takes on a universal character. The reason for this is that movement occurs over time. To go back to Figure 4.1, the universal, dog, stands to individual breeds of dogs (poodles, collies, boxers, etc.) as the movement of an object stands to the object conceived as being in a series of states at a series of times (like a strip of movie film). The classification of movements would be like the classification of a group of film strips. The proper way to do this, according to Aristotle, is in terms of ends—in cases of locomotion, natural ends.[8]

You would put all strips showing people laying the foundation of a house under the label, "The process of laying a foundation." If a film strip showed a completed foundation in the last frame, it would be easy to classify. All such strips would go together regardless of what the previous frames contained. There could have been many or few builders; the process could have been more or less mechanized; the materials might be stone, concrete, or wood. The builders may have had to make elaborate preparations (as with cantilevered houses) or just make marks on a flat rock. In other words, the efficient causes may be various but the final cause is the same.

Aristotle's classification of movements in terms of final rather than efficient causes corresponds (as behaviorists will have already noted) to

Skinner's conception of an *operant* as a class of movements with a common end. A rat's lever press is an operant because any movement of the rat that eventuates in a lever press (actually the closing of an electrical switch attached to the lever) is an instance of the same operant regardless of its efficient cause. The rat could hit the lever with its tail, its nose, its paw; all would be repetitions of the same operant. Of course, it is required that the rat press the lever. A current of air in the chamber that succeeded in pressing the lever would not fall into the same class. The rat for Skinner is like matter for Aristotle: necessary for the occurrence of an operant, but less important than the end of the movement (the lever press). This Aristotelian conception, the *operant,* is, in the opinion of many behaviorists (myself included), Skinner's major contribution to psychology. It shifts the focus of behavioral investigation away from efficient causes, which in the case of such movements as lever presses could only be conceived in physiological, cognitive, or mentalistic terms, and toward final causes— contingencies of reinforcement.

For Aristotle, while complete understanding requires knowledge of both efficient and final causes, final causes alone are sufficient to classify the movement. Understanding of final causes therefore naturally comes first, *before* understanding of efficient causes. Teleological behaviorism, as defined in Chapter 2, is therefore a true psychological science. The answer to the question, What does all this have to do with concepts such as *belief* and other "propositional attitudes" of modern philosophy of psychology? must wait for the discussion of Aristotle's psychology in Chapter 5.

Notes

1. Unless otherwise indicated, all quotations of Aristotle in this book are taken from Richard McKeon (1941).

2. In an essay entitled, "Shame, Separateness, and Political Unity: Aristotle's Criticism of Plato," Martha Nussbaum (1980) traces Plato's distinction between opinion and account to (a largely behavioral) definition of shame common to Plato and Aristotle. According to Nussbaum, the two philosophers agree that shameful behavior consists of failure to live a life dominated by reason together with knowledge of such failure. Plato's political remedy for shameful behavior as outlined in *The Republic,* Nussbaum argues, is to impose reason from without by authoritarian rule. Aristotle's remedy is to encourage individual reason by endowing citizens with maximal autonomy in action. The distinction is analogous to modern authoritarian versus democratic political systems.

Nussbaum takes *The Republic* as a serious proposal for a real society. There may well be external evidence (from the *Laws,* for instance) that Plato did indeed intend *The Republic* to be a model for an actual society. Within the dialogue itself, however, the ideal society is proposed solely as a model (writ large) of individual behavior. In modern times, the analogy between *The Republic* and a real society is more like the analogy of a Turing machine to a computer (an abstraction not necessarily to be constructed) than a serious proposal. In either case, however, Plato's disturbing implication remains that the better society became as a whole the less need there would be for individual reason.

3. Only deductive reasoning from true premises can serve as "the account" that both Plato and Aristotle agree must accompany true opinion to create true (unqualified scientific) knowledge. For Aristotle, the only areas of knowledge that may be strict sciences are geometry, arithmetic, and optics because "no attribute can be demonstrated nor known by strictly scientific knowledge to inhere in perishable things" (*PA*, I, chap. 8, 75b, 24). Other areas of knowledge (including psychology) can only approach these three.

4. Thus, in writing the script, the author would be, according to Plato, conforming well or poorly (depending on his or her personal characteristics) to a pattern already there in the world. The actors in acting and even the viewers in viewing would be doing the same (and not necessarily at a lower level; for actors the script and for viewers the movie are only *parts* of the world reflected by their acting and viewing).

5. I shall discuss later what Aristotle means when he says that something is "in" something else.

6. Despite their names, "material" and "formal" causes are relevant to a *formal* analysis (the middle set of circles). Material and formal causes get their names from matter and form but their application is in the world of description, not in the world (of matter and form) itself (where efficient and final causes apply).

7. Renaissance physics moved in the opposite direction, embracing all change as forms of efficiently caused locomotion. Unquestionably this revolution was good for physics. But looking at psychology today one may question whether it was good for psychology.

Kuhn further points out that Aristotle's denial of the existence of a vacuum (a void) is consistent with his notion of place as an attribute (like redness or roundness). A vacuum would be like an attribute without a subject (the smile without the Cheshire cat), a logical impossibility. In functional terms, a vacuum would be like form without matter, Aristotle's objections to which, we have already discussed.

8. It is aside from my purpose here to discuss forms of change (coming into being and passing away) that are not considered by Aristotle to be individual movements, but it is worth noting here that a thing has being (is a thing) to the extent that it takes part in a definable movement (in its wide, Aristotelian, sense). A thing comes into being, therefore, when a new movement comes into being. A movement may come into being by splitting off from another movement (much as we conceive a species coming into being in evolution). Both the physicist (hence, the psychologist) and the philosopher are properly students of coming into being, but the physicist studies the process as *movement* (functional analysis) while the philosopher studies the process as *being* (formal analysis) (*M*, XI, chap. 3).

5

Aristotle's Psychology and Ethics

De Anima (On the soul) is Aristotle's analysis of the behavior of organisms. Yet at the same time it seems to be an exercise in mentalism, a discussion of sensation, perception, imagination and thought. How can an analysis of behavior be the same thing as a study of mind? This chapter, the core of the present book, attempts to show that these two endeavors, the analysis of behavior and the study of mind, as Aristotle conceived them, are one and the same. But first it is necessary to answer four questions: What (according to Aristotle) is the soul? Where is the soul? What is the relationship of the soul to the body? How can the soul be understood?

Aristotle preserves the tripartite character of Plato's division of the soul; for Aristotle, however, the division is not along moral, but more immediately functional, lines. The three parts are nutritive, sensible, and rational. All living organisms function nutritively (have a nutritive soul); all living animals function sensibly (have a sensible soul); all living humans function rationally (have a rational soul). Since humans have sensible and nutritive as well rational souls, they are animals and organisms. Aristotle draws an analogy between the parts of the soul and geometrical figures (*DA*, II, chap. 3, 414b, 28). Just as it is impossible to draw a square without a potential triangle within it (formed by the diagonal and two sides), it is impossible for a person to function rationally without sensation and impossible to function sensibly without nutrition. Each higher function is built upon a lower function, and in the exercise of the higher the lower is also exercised. Another analogy might be between a theme, a movement of a symphony, and the symphony as a whole.

What Is the Soul?

The soul of an organism, for Aristotle, is the pattern of its movements ("movements" defined broadly as in Chapter 4). *De Anima (DA)* classifies and analyzes such patterns. As Chapter 4 indicates (see the rightmost circles of Figure 4.2) the components of a functional analysis are matter and form. When the "substance" being analyzed is the movement of an organism, its soul is the form of that movement and its body is the matter.

Form, for Aristotle, as for Plato, is not just the shape of something but its classification, its niche (or the bin in our imaginary museum into which it goes.) The movements of a hunter, for instance, would be classified, not according to their paths in space, but according to their nearer purposes (killing a deer, trapping a fox, and so forth) and in turn to further purposes (obtaining food, clothing, and so forth). The souls of organisms, there-fore, are classifications of movements of organisms according to their purposes (*teleological* classifications).

Where Is the Soul?

In *Physics* (IV, chap. 3) Aristotle enumerates eight conceptions of *in*, in-cluding our common conception: as water is *in* a jar. The soul is *in* the body, Aristotle makes clear (*DA*, II, chap. 2), as health may be in the body. It is a mode of functioning of the body. Later Aristotle will say that knowledge is *in* the soul in the same sense that the soul is *in* the body (and that health is *in* the body). Once Aristotle's functional (and formal) con-ceptions of what it means for something to be within something else are understood, his conception of the relationship of knowledge to the soul (to be discussed later) becomes easier to grasp.

What Is the Relationship of the Soul to the Body?

In the *Metaphysics (M,* VII) Aristotle says that the form of a circle is easy to conceive independent of its matter (say, a piece of bronze) because you can have circles made of any substance—wood, stone, and so forth. Since circularity may be so easily transferred from one kind of matter to another, and is found in all sorts of natural objects, we easily conceive of circularity as residing essentially in the form but only incidentally in the matter.

The case of the human soul, Aristotle claims, is the same, but in this case it is more difficult "to perform the abstraction." Our souls reside essentially in our behavior, not in our "flesh and bones." We have trouble understanding this fact because we do not find, in the natural world, other material objects, the behavior of which could be classified in terms of human goals. Because human souls are found in nature only in human bodies, claims Aristotle, we tend to particularize the human soul in the human body. If there were other material objects (perhaps like Dolly II of Chapter 1) whose behavior could be classified in terms of human goals, we would see much better how incidental our bodies are to our souls.

How Can the Soul Be Understood?

Aristotle, like Plato, does not even consider that a person could know his
or her own soul by direct (introspective) observation. He says at the outset
(*DA,* I, chap. 1, 402a, 10): "To attain any assumed knowledge about the
soul is one of the most difficult things in the world." The reason such
knowledge is difficult is not because the soul is hidden inside of us but
because the movements of human beings are so complicated. The contemporary writer Italo Calvino (1988) quotes Hugo von Hofmansthal (1874–
1929; the literary critic and librettist for Richard Strauss) as follows:
"Depth is hidden. Where? On the surface." In the same sense, the soul is
hidden. Understanding the soul, therefore, "is one of the most difficult
things in the world," in the sense that understanding an abstract painting
may be difficult.

Just as the "depth" of an abstract painting is not to be found behind
the painting, so too, according to Aristotle, the depth of the human soul is
not to be found behind behavior—in the body of the behaving person.
Linguists such as Noam Chomsky (1980) are therefore consistent with
Aristotle when they seek to discover "deep structure" in human (verbal)
behavior. They are inconsistent with Aristotle, however, in supposing that
the deep structure they find must be coherently represented in an "organ"
within the human body. Looking for such an organ would be for Aristotle
(and Hofmansthal) like looking for the deep structure of an abstract painting in the room behind the wall on which it is hung. *De Anima* is an
attempt to reveal the soul, not by digging deep but by taking a step back
and perceiving (in its most abstract sense) the behavior of organisms
(where behavior includes qualitative change as well as locomotion) as they
function in their natural environments.

In other words, knowledge of the soul is direct, for Aristotle, only to
the extent that an overt pattern of movement may be perceived as such.
Aristotle persists in this "order of investigation" throughout *De Anima* as
it proceeds from the (lowest) powers of nutrition to the (highest) powers
of reason. People, no matter how wise or insightful, are never supposed by
Aristotle to be able to know the soul either of other people or of themselves in any other way. This way of reading Aristotle is important; if valid,
it means that, like Plato, Aristotle rejects (essentially) private knowledge.
Recall Irwin's (1980) statement quoted in Chapter 2: "Aristotle has no
reason to think that psychic states—perceptions, beliefs, desires—must be
transparently accessible to the subject, and to him alone. Even if there are
such states, this feature of them is not the feature that makes them psychic
states. Psychic states, for human souls as for others, are those that are
causally relevant to a teleological explanation of the movements of a living
organism" (p. 43).

When Aristotle identifies the soul with the functioning of the body,
he is referring to the body as a whole, not its parts. For instance, the
mechanisms inside an automobile (carburetor, steering mechanism, mo-

tor, brakes, and so forth) all must work for the automobile to work. However, the functions of the automobile as a whole (its cornering, its accelerating, its comfort, its stopping distance, and so forth) and not the mechanisms inside it would be, if an automobile could move itself, its soul. To know the soul is thus, for Aristotle, to understand how the body as a whole functions, as an expert driver would know a car, not how the parts of it work, as an expert mechanic would know a car. (Complete understanding of the living animal, however, would require both kinds of knowledge.)

States of Mind

A state of mind is a power to act in a certain way. Cognitivists, physiologists, and mentalists will view mental states or powers as interior states or perhaps dispositions based on interior structural capacities (states of an internal computation mechanism, physiological states, or transcendental states divorced from any physical object). Behaviorists must, to be consistent, view such states as patterns of overt behavior, as patterns (however complex) of observable change. Observable change need not be only locomotion. For instance, a person's face may redden without any locomotion of the person. But mental states are properly identified by the behaviorist not with such qualitative changes in isolation (nor with covert or hidden brief locomotions) but with temporally extended patterns of observable overt behavior (which might include patterns of observable qualitative change). When mental states are obscure, their obscurity must, for the behaviorist, be due to complexity in patterns of overt events.[1]

It is significant that Greek terms others translate as "state" or "power" are often translated by Randall (1960, pp. 254 and 267, for instance) as "habit." Habits may be obvious or they may be difficult to detect because of subtlety of pattern. A person's transition from a social drinker to an alcoholic, for instance, may escape the observation of the person herself as well as of friends and relatives, not (necessarily) because of any unobserved internal events but because of unperceived differences between previous and current patterns of behavior. Similarly, movement from sickness to health or from ignorance to knowledge may be seen as a change in the pattern of overt behavior rather than cognitive, physiological, or spiritual change.

Potentiality and Actuality

Aristotle begins the second book of *De Anima* (II, chap. 1, 412a) with a discussion of what it means to be alive.[2] To be a living natural body (an organism) means to have both a body and a soul. In developmental terms the relationship of body to soul is that of potentiality to actuality. A body by itself is only potentially alive. A body with a soul is actually alive.

Aristotle says (*DA,* II, chap. 1, 412a, 20): "The soul must be a substance in the sense of the form of a natural body having life potentially within it. But substance [in the sense of form] is actuality, and thus soul is the actuality of a body as above characterized."

Aristotle goes on to distinguish between two degrees of actuality. The first degree of actuality is the condition of being in the midst of a complex pattern of behavior but not necessarily doing anything specific at a given moment. For instance, once a performance of a play begins, the play is actually being performed even, in a sense, between acts. Similarly, an orchestra is in a sense actually playing Beethoven's Ninth Symphony even at the moment between movements when it is actually (in another sense) doing nothing. The soul is identified by Aristotle with the first degree of actuality. He compares this sort of actuality to "knowledge possessed" but not necessarily employed at this very moment. "For both sleeping and waking presuppose the existence of soul, and of these waking corresponds to actual knowing [knowledge possessed and employed], sleeping to knowledge possessed but not employed."

In addition to the two degrees of actuality there are two kinds of potentiality. The first is like the potentiality of a boy to be a general; the second, like the potentiality of a man to be a general. Aristotle traces these distinctions with several examples (a tree, an eye, an axe, an animal).

In the case of a human being, sperm and egg would be potentially human in the very first sense (like the separated blade and handle of an axe) while an adult body is potentially human in a higher sense. Its first degree of actuality (its soul) is its power of functioning as a human being, which is possessed during sleep. If a body should lose this power, it thereby loses its soul. The second degree of actuality is actually engaging in behavior characteristic of a human being—that is, acting rationally—which cannot be done while asleep. Figure 5.1 illustrates these relationships.

The relationship of the first degree of actuality to the second degree of actuality is that of the universal to the particular as this relationship is applied to movement. A person's soul (his overall pattern of movement) is the person's first degree of actuality. The movements themselves are the second degree of actuality. Particular states of the soul are modes of movement that fit into the larger pattern. Aristotle says that the state of "anger should be defined as a certain mode of movement" (*DA,* I, chap. 403a, 25). A particular act such as striking a blow or turning red in the face is actual only the second sense. In more modern terms, the first degree of actuality is the *context* of the second. A pianist playing a sonata, for instance, is actually (in the first degree) playing the sonata. The sonata as a whole forms the context of the actual (in the second degree) playing of each individual note. This relationship is important because it underlies Aristotle's conceptions of both sensation and imagination, to be discussed later.

The direction of development in time goes from the remotest kind of

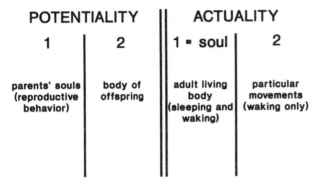

POTENTIALITY || ACTUALITY

1 | 2 || 1 = soul | 2

parents' souls (reproductive behavior) | body of offspring || adult living body (sleeping and waking) | particular movements (waking only)

--›DIRECTION OF DEVELOPMENT--›

‹--DIRECTION OF INVESTIGATION AND UNDERSTANDING‹--

Figure 5.1 Degrees of potentiality and actuality in the development and understanding of the soul.

potentiality to the nearest, then the first degree of actuality, then the second degree. The pieces of an axe are made first, then the axe is put together, than it is used. Its potentiality to be used must come before its use—at least conceptually, although Aristotle acknowledges that in real time both may appear at once. In the case of human beings, the ability to reason, say, must precede reasoning itself.

However, Aristotle distinguishes between the order of development of a living organism over time and the order in which we come to know that organism. We come to know a living organism in exactly the opposite order from its development. First we perceive its individual movements, then we perceive its pattern of movement; that is as "deep" as our perception goes. The highest order pattern *is* its soul. Only at that point, by scientific demonstration, can we reason backward to potentialities. As Aristotle says (*DA,* II, chap. 4, 415a, 19): "In the order of investigation the question of what an agent does precedes the question, what enables it to do what it does." Then Aristotle takes the issue still further. He goes on: "If this is correct, we must on the same ground go yet another step farther back and have some clear view of the objects of each; thus we must *start* with these objects, e.g., with food, with what is perceptible, or with what is intelligible" (italics in original).

The Nutritive Soul

True to his stated plan of beginning all investigations of the soul with its "objects," Aristotle's discussion of nutrition begins with what he conceives to be the most fundamental object of all living things: "to partake in what is eternal." Since no living thing "can for ever remain one and the same, it tries to achieve that end in the only way possible to it" (*DA,* II, chap. 4, 415b, 4), which is by reproductive movement.

It is in the context of reproduction that Aristotle discusses the soul as the cause of the living body. Remember, for Aristotle, a cause must be the middle term in an argument. In Figures 4.1 and 4.2, a cause is represented by the middle circle. In Aristotle's functional analysis the inner circle is the body, the middle circle is the soul, and the outer circle is the "living body" (the organism). Aristotle says: "The soul is the cause or source of the living body. The terms cause and source have many senses. But the soul is the cause of its body alike in all three senses which we explicitly recognize. It is (a) the source or origin of movement, it is (b) the end, it is (c) the [formulable] essence of the whole living body."

The concept of the soul as "the source or origin of movement" does not mean that it can be an immediate efficient cause of movement as even the very best modern psychology textbooks imply (e.g., Watson & Evans, 1991, p. 78). Rather, by their reproductive movements, the nutritive souls of organisms (plant, animal, or human) are conceived as "(a) the source or origin of movement" of their offspring. Alternatively, the souls of the offspring may be conceived as "(b) the end" of the reproductive movements of their parents. Third, the soul is "(c) the essence of the whole living body" in the sense of "an unbroken current of the same specific life flowing through a discontinuous series of individual beings" (parents and offspring), as Smith indicates in a footnote to his translation of *De Anima* (McKeon, 1941, p. 561). None of the three senses in which Aristotle conceives the soul to be a cause corresponds in any way to the (Cartesian) concept of an exclusively human soul (located inside a body) efficiently causing each individual voluntary act.

Reproduction is more fundamental than nutrition because the latter may be done for the sake of the former, but not vice versa. But both reproduction and nutrition are functions of the nutritive soul. The object of nutrition is food. In the act of eating, the form of the digested food changes to that of the eater. Food contributes to a person's movements (the person's form) only as an intermediary in a chain of efficient causes (like the rudder steers a ship as an intermediary between a hand and the ship itself). The organism takes on the matter of the food (but not its form). The eater is thereby increased in bulk. By changing its form, the food maintains the being of the eater and thereby becomes an agent in reproduction. Thus, Aristotle is careful to build the analysis of the subsidiary function (nutrition) upon the more fundamental (reproduction). This pattern is maintained in his analysis of sensation and reason.

Sensation

In nutrition, according to Aristotle, the body takes in the matter of the object (food) without taking on the form, whereas in sensation the soul takes on the form of the object without taking in the matter. But Aristotle's concept of the soul's taking on the form of an object does not correspond to the modern cognitive conception (as illustrated in Chapter 2) wherein an object is internally represented for further processing. In the modern conception, information enters the nervous system from outside. After sensation the system is in a different state from before. It possesses something it did not have before: the image or the representation of the sensed object. For Aristotle, however, sensation is not a change of state, where something (an image) is acquired that was not there before, but actual movement.

The difference between the pattern of a person's actions (the person's soul) before sensation and afterward is that afterward any complete description of the person's actions must contain a description of aspects of the object (its sensory qualities). "What has the power of sensation [a living animal] is potentially like what the perceived object is actually; that is, while at the beginning of the process of its being acted upon the two interacting factors [the behavior of the animal and the qualities of the sensible object] are dissimilar, at the end the one acted upon [the behavior of the animal] is assimilated to the other [the color, smell, sound, etc., of the object] and is identical in quality with it" (*DA*, II, chap. 5, 418a, 5). The process Aristotle is describing is behavioral discrimination. Objects may be classified according to their colors, say. The animal's discrimination enables an observer of the animal to classify its behavior in the same way (by the colors of objects).

A *discrimination* as conceived in modern behaviorism (Rachlin, 1991), is a pattern of events in time, not a particular event. A driver's discrimination of red traffic lights from green traffic lights, for instance, requires at least two specific events, going on green and stopping on red. (An additional two events, not-going on red and not-stopping on green may also be required depending on circumstances.) Furthermore, because no discrimination is perfect, the events are probabilities (in a contingency table) each itself consisting of a series of observations (see Chapter 2). A discrimination is really a statistically significant tendency (to stop on red and go on green) rather than any absolute obedience to discriminative stimuli (traffic lights). In cases where stimuli and responses are graded (for instance, speed of the car and pressure on the accelerator) a discrimination is a correlation rather than a contingency.

This whole pattern forms the background (the context) for each individual discriminative act (just as the whole football team forms the context for each player). In Aristotelian terms, the discrimination as a whole is actual in the first degree (is part of the soul) while each individual act of discrimination (each instance of stopping at red or going at green) is actual only in the second degree. Any individual act is meaningful and may be called a "sensation" (of red) only in the context of the discrimination as a whole. A person who stops at a red light just once, for instance, may have run out of gas, may have reached her destination, or may be actually stopping for the red light; it depends on the context of the act. And context consists of the other acts in the pattern, not events that may or may not occur or have occurred in the person's nervous system or consciousness.

If a driver generally stops at red lights and goes at green lights, the driver may therefore be said to discriminate between red and green. Given that this discrimination (and other discriminations) between red and green do generally occur, then, in a particular instance, when the driver actually does stop at a red light, we may say that the driver sensed the color of the light. We can then classify that bit of behavior under that particular discrimination. Doing so (making the classification) is an act (*our* act) of matching behavior and color. Stopping at red, saying "red," underlining things in red, etc., are discriminations of the same sort. The investigation and analysis of sensation is the investigation and analysis of this process (of discrimination).

Because individual sensations are individual movements, it is not possible for the sensing animal to make an error. An observer may make an error by classifying a given movement wrongly (a driver might seem to stop at a red light but might actually be doing so incidentally, to get directions). But with respect to the objects of sensations, "no error is possible" (*DA*, II, chap. 5, 418a, 10). It cannot be an error to behave in a certain way in the presence of a given object. If an animal discriminates between red and not-red, for instance, the animal is sensing the color. If an animal does not make the discrimination, the animal is not sensing the color. There is no room in Aristotle's way of thinking for an animal that senses the color yet senses it wrongly. Any discrimination must be correct, precisely because it *is* a discrimination. It is only when those movements constituting the discrimination come to be classified with others by an observer (including the animal itself as an observer of its own behavior) that an error is possible.

This issue is important because many other thinkers have adopted Aristotle's supposition that sensations are free from error. However, for Aristotle, the reason is that the term "error" applies only to inconsistency between sensations. This differs from the reasoning of many subsequent thinkers, that sensations are absolutely certain because they are so close to some internal perceiver. For Aristotle, sensations are free from error in the (logical) sense that a person who draws the target around the point where his shot lands cannot miss the target, not in the (empirical) sense that a person cannot miss a very close target.

Particular sensations are particular movements defined according to their objects. Seeing, for instance, is a movement (of a whole person) affected by its object: color. Any bodily movement that discriminates among colors is an instance of seeing. However, according to Aristotle, colors cannot by themselves affect movements of whole animal bodies. Instead, colors influence bodily movements through another form of movement: light.

Thus, there are two kinds of movement involved in seeing. First there is seeing itself, a form of movement of which only whole living animals are capable. Second, there is light, also considered by Aristotle to be a movement. But for Aristotle light is a qualitative change involving no locomotion; in our terms, its speed is infinite. A body capable of being moved in this way is called transparent. According to modern physics, light travels from a source through space to an object, is reflected by the object, goes back through space, and finally reaches our eyes. According to Aristotle (*DA*, II, chap. 7) light is a qualitative change of the space between the object and our eyes, a transformation of that space from *potential* transparency to *actual* transparency. When that space serves as a medium between the color and the animal, it is actually transparent (has moved from darkness to light).

A medium is required because "if what has color is placed in immediate contact with the eye, it cannot be seen" (*DA*, II, chap. 7, 419a, 15). The modern equivalent of an Aristotelian medium in a physical (including biological) process would be a catalyst in a chemical process rather than a link in a machine. If A influences B through medium M, there is no point even conceptually at which M is affected by A but B is not yet affected. Rather, A affects B but M must be present for it to do so. The eye, as an organ, is necessary for sight, but the eye itself does not *see;* that is, the eye cannot make the sorts of (discriminatory) movements defined as seeing. The form of the eye's movement is that of light, not of seeing. Furthermore, (except when looking in a mirror) we do not see our eyes themselves, first because we cannot move with respect to our eyes as we would have to do in order to see them, second because we have no organs with which to see them.

Aristotle discusses each of the other particular senses (hearing, smell, taste, and touch) similarly in terms of its object, medium, and organ. The relationship between these three factors in sensation corresponds to that in other parts of Aristotle's functional analysis of change. Organ is to object as potential is to actual, as matter is to form (as particular to universal). In each case the medium is the cause of a transition from the former to the latter. Thus, for Aristotle, the organ actually becomes the object as a form or a pattern of movement. The influence of objects on organs (considered in isolation) is indeed (for Aristotle) efficient causation "conceived of as taking place in the way in which a piece of wax [the organ] takes on the impress of a signet-ring [the form of the object] without the iron or gold [the matter of the object]" (*DA*, II, chap. 12, 424a, 20). This efficiently caused movement within the eye, however, is distinct from another process: the influence of the object on the body as a whole. It is this larger process that Aristotle identifies as sensation.

At the end of his discussion of specific sensory modalities (*DA*, II, chap. 12, 424b), Aristotle makes this distinction clear, using smells as an example. In terms of Figure 2.2 a smell may be a discriminative stimulus (an S^D) signaling a given contingency; as an S^D a smell can affect only

beings capable of sensory discrimination—animals—not plants or inanimate objects. When an odorous substance does affect plants or inanimate
objects, it does so as a simple efficient cause, not as an S^D; hence not
(according to Aristotle) as a smell: "A smell is just what can be smelt, and
if it produces any effect it can only be so as to make something smell it."
Smelling here is understood as a movement of a whole animal. Nonethe-
less, odorous things also seem to affect objects without souls (physical
objects, to us) such as the air. To this, Aristotle answers, first, that since the
air has no boundaries the effect "disintegrates." Second, "smelling is more
than . . . [being affected] by what is odorous—*what* more? Is not the
answer that, while the air owing to the momentary duration of the action
upon it of what is odorous does itself become perceptible to the sense of
smell, smelling is an *observing* of the result produced" (italics in original).
Thus, the air is affected by what is odorous (as a tree may be split by "the
air which accompanies thunder") but not by a *smell* that (like the sound of
thunder) can only affect something with a sensible soul (that is, a whole
animal).

Perception

Aristotle's account of perception (*DA*, III, chaps. 1–3) is important because perception, for him, is the basis for all knowledge. After discussion
of the five special senses (sight, hearing, touch, taste, and smell) Aristotle
proceeds to the first level of complexity—the common sensibles—objects
that may be perceived by more than one sense (for example, movement,
rest, figure, magnitude, number, unity).

Common sense is the simplest type of perception (as distinct from
sensation). It is an elemental capacity, possessed by almost all animals, to
directly perceive certain movements of objects. The examples of movements perceptible by common sense cited by Aristotle are defined (as is
typical) in terms of ends. The common sensible, *rest,* for instance, is conceived, not as an absolute state, but as a coming to rest (relative to the
perceiver) of something that is moving. Movements are perceived through
one or more of the five (special) senses. Suppose you *see* something move.
If you had a special sense for movement you would then be perceiving
with one sense (sight) an object of another sense (that of movement).
Therefore, there cannot be a special sense for movement. "If that were so,"
Aristotle says, "our perception of it [an object seen to move] would be
exactly parallel to our present perception of what is sweet [one sense] by
vision [by another]" (*DA*, II, chap. 1, 425a, 22). Aristotle's rejection of a
special sense for movement implies a rejection of the possibility of an
internal sense for movement (an internal sense organ that stands to the
special sense organs as they do to their objects or their media) because
touch is already an internal sense (for which the skin is conceived as the
medium). If there were a special internal organ for common sense, the
objects of that organ would overlap with those of touch. Perception of

tactual movement by that organ would be like "perception of what is sweet by vision." Just as the special senses are conceived as movements of the whole organism, not as the functioning of an organ, common sense is also a movement of the whole organism and not the functioning of an internal organ.

The difference between special and common sense is in the complexity of the pattern discriminated in the *object,* not, as many interpreters (St. Thomas Aquinas, for instance) say, in the *type* of organ. Discrimination of color, for instance, is only possible with the eye. But the pattern that constitutes movement through space of a visible object may be also a property of a tangible object (as it moves along the skin) and is therefore a common sense. Again, the difference is in the object (hence in the pattern of discriminative behavior) not the organ. Since the same object is perceptible by different senses, the senses may disagree. Hence, errors of common sense are possible. Thus, you may hear an airplane moving to the left but, when you look, you see it as moving to the right. The discrimination (between left-moving and right-moving objects) must be in error with respect to one of the sense modalities.

A still more complex case of perception (more complex than common sense) is the discrimination between the objects of different senses, for example, white and sweet. The fact that this discrimination is possible implies that "what asserts this difference must be self-identical, [one thing] and as what asserts [is one thing], so also what thinks or perceives [must be one thing]" (*DA*, III, chap. 2, 426b, 22). [Otherwise,] "even if I perceived sweet and you perceived white the difference between them would be apparent" (*DA*, III, ch. 2, 426b, 18).

Here is the central problem of Greek philosophy—the distinction between the one and the many—brought down to its most elemental level. Just as in thought the number two is simultaneously one thing (the number) and two things (two units), so the simultaneous perception of white and sweet is one thing and two things. In the case of perception (and, as we shall see, in thought), both the unity and the diversity are movements. Movements cannot, it seems, be two things at once. Aristotle says: "it is impossible that what is self-identical [the one organism that must discriminate white from sweet] should be moved at one and the same time with contrary movements insofar as it [the organism] is undivided and in an undivided moment of time" (*DA*, III, chap. 2, 426b, 29). In other words, a particular instance of discrimination of white from nonwhite belongs in one discriminative pattern (one context) while a particular instance of discrimination of sweet from nonsweet belongs in another. To which pattern does a particular instance of discrimination of white from sweet belong? Because it is impossible (for the single organism that must discriminate white from sweet) to behave in contradictory ways at the same time, the particular instance of discrimination of white from sweet must belong to *both* patterns. How can this be?

Aristotle's solution is to view the discrimination between white and sweet as a single discriminative movement that takes part simultaneously in two separate patterns, "just as what is called a 'point' is, as being at once one and two, properly said to be divisible, so here that which discriminates is *qua* undivided one, and active in a single moment of time, while so far forth as it is divisible it twice over uses the same dot at one and the same time" (*DA,* III, chap. 2, 427a, 10). An example would be the point of tangency between two circles. Another example might be the playing in a symphony of a single note or phrase that is simultaneously part of two overlapping but different themes. Thus, a single discriminative act (the discrimination of white from sweet) is part of two patterns of acts (the discrimination of white from nonwhite and the discrimination of sweet from nonsweet).

Imagination

As we have said, according to Aristotle, perception of the special objects of sense is free from error but perception of the common sensibles is subject to error. A still more complex case, which Aristotle classifies as a form of thought, is what present-day psychology calls perception—the perception of objects as such or of complex attributes of objects.

If, for instance, a piece of paper is perceived as a white object, it is only incidentally a piece of paper. As with other special sensibles, perception of whiteness as such is not subject to error. However, when the incidental object is itself perceived—when the piece of paper is perceived as a piece of paper (not as just a white object)—the perception might be false, the white object might, for instance, be a shirt. The kind of movement in the perception of incidental objects that makes error possible is imagination.

Aristotle says, "imagination must be a movement resulting from an actual exercise of a power of sense" (*DA,* III, chap. 3, 429a, 1). That is, as far as the actual movements of the animal are concerned, imagination is the same as sensation. The difference is that the objects of sensation are present during sensation while during imagination the objects of imagination are not present. Aristotle does *not* say that the objects which are present in the world during sensation (which sensation discriminates among) are present inside the animal (as representations, internal images, neural discharges or anything else) during imagination. What he does say is that the movements the animal makes during imagination with the objects absent are the same as those the animal makes during sensation with the objects present, except in that one respect. If you generally behave one way in the presence of, and another way in the absence of, red lights, you are discriminating between red lights and other things. But if, on an occasion, you behave in the absence of a red light as you normally do in its presence, you are on that occasion imagining a red light.

Just as an animal may imagine the special objects of sense (simple colors, sounds, smells, and so forth), it may imagine the incidental objects of sense (people, shirts, chairs, lions, etc.). A person who generally behaves in one way in the presence of trucks and in another way in the absence of trucks is discriminating between trucks and nontrucks. A general pattern of this kind is necessary before that person can be said to perceive a truck. Given such a pattern, and the presence of a truck, a given "base assertion" (*DA*, III, chap. 7, 431a, 9) such as "There is a truck" would be considered by Aristotle as an individual perception of the truck. The same assertion in the absence of the truck is therefore an individual imagination of the truck. But neither individual perception nor individual imagination of a truck can occur except in the context of the overall discriminative pattern.

If you behave occasionally in the absence of a chair, a lion, a truck, as you normally would in the presence of those objects you are, on those occasions, Aristotle would say, imagining those objects. Again, this conception of imagination is entirely dependent on a prior conception of sensation or perception. Consider the following example. Suppose a driver typically drives in the left lane of the road. When a truck comes roaring up from behind, the driver's typical reaction is to swerve suddenly to the right. Each time this happens (in its typical form) swerving may be conceived as a discriminative act.[3] Given this habit, a driver who swerved in the absence of a truck would be imagining a truck—the object of the movement is missing. This is what Aristotle means when he says, "imagination is . . . impossible without sensation" (*DA,* III, chap. 3, 428b, 11).

Imagining is *acting* not dreaming; vividness of imagination is not vividness of interior image but of overt behavior. This extreme operational and behavioral conception is not only entirely consistent with Aristotle's discussion of imagination but also fits with his discussion of sensation and perception (as we have seen) and of thought (as we will see).

Suppose two people sitting in a room (Jack and Jill) are both asked to imagine a lion. Jack nods his head, closes his eyes, and says, "Yes, I see the lion, it has a mane and a long tail, it's brown, it's wandering around the room, now it's roaring." Jill runs screaming from the room. For Aristotle, Jill would be truly imagining the lion. Jack, on the other hand, is imagining not a lion but perhaps a picture of a lion or a movie of a lion or a story of a lion.

Consider another borderline case between perception and imagination. Suppose you are standing near a road and a bus goes by. You follow the bus with your eyes until it is just a dot on the horizon. Some friends come along. You point to the dot and say, "See that bus?" They say, "No, you're just imagining it." Are you seeing it or just imagining it (or perhaps seeing the dot but imagining the bus)? According to Aristotle you are actually seeing the bus because the bus is there. If it were not there (if, say, you had blinked and the bus were actually out of sight; the dot could be a bump in the road) your friends are right and you are just imagining the bus. The important point is that it is the object and only the object that makes the difference for Aristotle between perception and imagination. Your behavior is the same in either case. In deciding the issue between you and your friends, the relevant question is: Is that dot a bus or is it not? The location, intensity, orientation, or even the existence of an image in your head would be entirely irrelevant.

In the above example, you as the observer would not know whether the object of your sensation was there or not—whether you are actually sensing it or just imagining it. But an actor on a stage knows that the object of his anger (say) is not there. He is just (temporarily) behaving as if it were, which is another way of saying he is imagining that object. A good imagination is not just an aid or a tool in good acting. Rather, for Aristotle, good acting *is* good imagining.

Thought

Thinking differs from perceiving and sensing not only by the complexity of the movement involved in thought but also with respect to the differing role played by imagination in the different processes. Because the objects of both sensation and perception must be present during the processes themselves, sensation has no meaning without the object and is incompatible with imagination, while perception, to the extent that it involves imagination, is wrong. Thought, on the other hand, requires imagination. "The soul never thinks without an image," Aristotle says (*DA*, III, chap. 7, 431a, 15).

In what sense does thought differ from sensation and perception so that imagination, so destructive of the first two processes, becomes absolutely necessary for the third? In thinking, alone of the three processes, the object is absent before the process occurs. Thus, thinking cannot get off the ground, so to speak, without imagination. A thought must begin with a discriminative act in the absence of its object.

Prior to thinking, the direct object of thought may be conceived to be present in the *potential* behavior of the thinker. "What it thinks must be in it just as characters may be said to be on a writing tablet on which as yet nothing actually stands written" (*DA*, III, chap. 4, 430a, 2). Once the thought has occurred, however, its object is present as an object of perception for the thinker as well as for other people. (Thus, our own thoughts are accessible to us in the same way as they are to other people—by observation of our overt behavior.)

A man doodling on a piece of paper and a woman proving a theorem in geometry may begin with the same discriminative action—an act of imagination. The difference between them is that the doodler continues to imagine while the geometer perceives her imagination as an object in itself; she perceives the abstract qualities of what she has imagined and uses them to extend her imagination. The process of thinking thus consists of an alternating series of imaginations and perceptions.

Perhaps the best modern illustration of the relationship between imagination and thought is a series of experiments by Walter Mischel and colleagues (summarized by Mischel, Shoda, & Rodriguez, 1989) on "delay of gratification." In these experiments, four-year-old children could obtain a preferred but delayed reward (a marshmallow) by *waiting* for the experimenter to come into the room by herself. Alternatively, they could obtain a less preferred reward (a pretzel) immediately by ringing a bell (which summoned the experimenter to the room). Children who waited with the pretzel exposed in front of them rang the bell significantly sooner than children who waited with the pretzel covered. Covering the pretzel presumably made it easier for the children to imagine that the pretzel was not there. Video recordings of the children waiting with a less preferred reward exposed illustrate how difficult their imaginative task was. They fidgeted, played games, made faces, closed their eyes, turned their backs to the exposed reward, or even tried to go to sleep. The connection between ability to imagine and ability to think is underlined by the startling finding that children who managed to wait longer with the reward exposed scored significantly higher than children who did not wait on SAT (college entrance) tests more than a dozen years later.

Further evidence that imagination played a significant role in this task is the reversal obtained with explicit instructions to imagine. With the pretzel exposed, children who had been told to imagine the pretzel was a toy log waited longer than children with the pretzel covered, who had been told to imagine how crunchy and tasty it was.

The cognitive interpretation of these results (Mischel's own interpretation) is that the instructions to imagine changed the child's internal representation of the pretzel, which then influenced behavior. For Aristotle, however, the child's waiting, conceived as behaving as if the pretzel was not there, is imagination, but the very same waiting behavior conceived as a means of obtaining the marshmallow rather than the pretzel is thought.

Aristotle conceives of thinking with no other object but itself as a rare form of behavior even among humans (I shall discuss this kind of thought in the next section). In ordinary cases, however, thought is practical: It has an object other than itself. Thus an act of practical thought is an act of pursuit of the good or avoidance of the bad. The good and the bad themselves are in the first place objects of appetite broadly conceived. Appetite is the definition of an end (right or wrong) that is (a) not present at the moment and (b) "capable of being brought into being by action" (*DA,* III, chap. 10, 433a, 17).

Appetite in turn may arise from desire. To "arise from," however, does not mean to "be efficiently caused by." Appetite arises from desire, for Aristotle, in the sense that a broader function (a wider category) arises from a narrower one. When appetite does arise from desire, the movement it causes will be pursuit of pleasure (or avoidance of pain). Pleasure and pain may or may not coincide with the good or the bad. Detailed discussion of differences between the good and the pleasurable and the bad and the painful is reserved by Aristotle for works on ethics, to which I shall turn next. But first let me summarize Aristotle's conception of thought.

Thought is an entirely behavioral process that, like other processes, may be understood primarily in terms of its ends. To understand an animal's thought, observers of behavior (whether that of another person or their own) first must perceive an end in the behavior as a whole. Then the behavior may be analyzed into acts of appetite, perception, and imagination. Imagination, in turn, may be analyzed into acts of memory and pure imagination. As in the analysis of movement in general (from the building of a house to the falling of a single stone), Aristotle always proceeds from overall functions to subsidiary functions.

In the movie *Mystery of Picasso* (made when Picasso was alive) the narrator begins by declaring that the movie's intent is to explore Picasso's mind. This is accomplished not by an interview or the elicitation of introspections but with few words and with hardly any pictures of Picasso himself. Instead, the camera focuses on the great painter's paintings as they come into being. Each one begins with a few lines or a simple design and evolves through many transformations into a final state. At one point Picasso's voice is heard saying, "Now I know what I was trying to do." Here is a case where viewers themselves can perceive the pattern of the painter's thought, which clearly consists of alternations of imagination and perception.

Imagination ordinarily requires memory; again, not in the sense that memory serves up images out of an internal storehouse of some kind, but in the sense that behavior is directed by objects previously but not currently present. If Picasso draws a cow present in a field, he is, according to Aristotle, perceiving the cow. If he draws a cow in the studio (where no cow is present), he is imagining the cow. Since Picasso had seen cows in the past, his imagination must have been based on memory. To many readers this will seem nothing more than the use of words to cover up ignorance (of the fundamental internal memorial process). But to Aristotle, the function of the concept of memory (like other mental concepts) was not to uncover internal processes but to help classify (hence to understand) *overt* behavior.[4]

Mentalistic terms—sensation, perception, imagination, memory, thought—must have been used just as vaguely and sloppily in Aristotle's time as they are today. *De Anima* is an attempt to give these terms specific meanings as categorizations of animal and human behavior not simply to reflect everyday use but to improve everyday use. If the behavioristic interpretation offered here of these mentalistic concepts—of the mind itself—fails to reflect Aristotle's true conception, it may well err by not being behavioristic enough. Thought, for instance, is not introduced by Aristotle, as it was here, as a process involving perception, imagination, and memory but (as is typical) in terms of its objects in everyday life: achieving pleasure or the good and avoiding pain or the bad, which in behavioristic terms are called reinforcers and punishers. These ends are the unmoved movers of animals in their everyday life. We turn to them now.

Ethics

Thomas Nagel (1980) in an essay on Aristotle's two works on ethics (the
Nicomachean Ethics [NE], and the *Eudemian Ethics*) makes an important
distinction, worth emphasizing, between Aristotle's concepts of sensation
and reason. Figure 5.2a (derived from Nagel's discussion but not drawn by
him) shows a cognitive conception of the workings of the sensible soul. In
Figure 5.2a the nutritive soul (N) is drawn separately from the sensible
soul (S′); their interaction is represented by the two arrows. The joint
action of these two separately conceivable but (according to Aristotle)
functionally inseparable parts of the soul is represented as S*. The souls of
nonhuman animals contain only these two parts. But, according to Aristotle, as the sensible soul functions, the parts are inseparable. For a giraffe,
for instance, S′ alone is meaningless. In behavioral terms, all of the giraffe's
behavior is explainable as S*, which is in turn conceivable as S′ at the
service of the more primitive N. However, no fraction of the giraffe's
behavior—no act, however primitive, of the whole giraffe—is explainable
in terms of N or S′ alone.

On the other hand, for a human with a rational soul, which stands to the
sensible part as the sensible does to the nutritive, R′ is not meaningless. In
Figure 5.2b the combined actions of the rational and sensible souls is represented as R*. R* is human life—human functioning. By far the largest
part of Aristotle's ethical writings is devoted to R*—practical reason—the
joint functioning of R′ (reason) and S* (appetite). However, for a human
being, according to Aristotle, R′ does have meaning outside of its functioning in R*. Some human behavior is not explainable as R′ interacting
with (and basically in the service of) S* but of R′ abstracted out of
its normal role as the form of human life and applied to "pure contemplation" of the abstract qualities of the universe as a whole—theoretical or
scientific reason.[5] As Nagel says (pp. 12–13), "Aristotle believes . . . that a
human life is not important enough for humans to spend their lives on. A
person should seek to transcend not only his individual practical concerns
but also those of society or humanity as a whole. . . . Comprehensive
human good isn't everything and should not be the main human goal. We

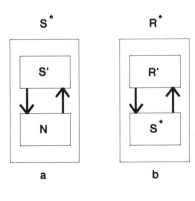

Figure 5.2 A cognitive
conception of the interaction
of the nutritive, sensible, and
rational souls.

must identify with the highest part of ourselves rather than with the whole."

If ethics is nominally Plato's main concern and epistemology is nominally Aristotle's main concern, it is not surprising that morality for Aristotle has about the same status as knowledge for Plato. Both are combinations of the good and the rational. Plato had said that knowledge is a combination of true opinion with an account. Having true opinion for Plato is the same as behaving well—being good. But good behavior without an account is like being blind but going the right way by accident (*Republic* VI, 506c). The account is necessary for knowledge, but the good behavior is essential in itself. Aristotle has a similar view of the role of natural virtue and reason in goodness (R* in Figure 5.2b). A person with natural virtue (bravery, for instance) but without reason may be led astray by the very virtuous qualities he possesses; without reason virtuous qualities may be expressed as excesses (bravery as rashness, for instance). It is reason (R') that steers these qualities to the mean between too much (rashness) and too little (cowardice). The visual imagery that Aristotle uses to express the function of reason as it guides natural virtue is almost exactly parallel to Plato's: "While one may be led astray by them [natural virtues], as a strong body which moves without sight may stumble badly because of its lack of sight, still, if a man once acquires reason, that makes a difference in action; and his state, while still like what it was, will then be virtue in the strict sense" (*NE*, VI, chap. 13, 1144b, 11). The difference is one of emphasis. For Plato, the good is more important than the merely rational. Rationality alone, for Plato, is nothing more than an "account," a rationalization—a driver of a chariot without horses. For Aristotle, on the other hand, the rational is more important than the merely good ("both children and brutes have the natural dispositions to these [good] qualities" (*NE*, VI, chap. 13, 1144b, 9). Goodness without reason, for Aristotle, is too easily subverted. For Aristotle, having natural virtue alone is nothing more than having healthy appetites. Again, the difference between Plato and Aristotle on this point is in emphasis rather than substance. Plato emphasizes the uselessness of reason without goodness but claims (*Republic* IV) that in practice the Good is achievable only through reason; Aristotle emphasizes the fragility of goodness without reason but claims that in practice pure reason is achievable only through ethical behavior.

The bulk of the *Nicomachean Ethics* is given over to a discussion of how reason (R') combines with appetite (S*) to produce practical wisdom (R*). I shall summarize this discussion presently. But first, following Aristotle's pattern, I turn to the ultimate object of an ethical life: human happiness.[6]

Happiness

Take any book of quotations (such as, Evans, 1968) and look up the topic, "happiness." You will find that happiness is viewed as an (unsustainable) emotion by about half of the authors (the pessimists). H. L. Mencken, for instance: "Happiness is probably only a passing accident. For a moment or two the organism is irritated so little that it is not conscious of it; for the duration of that moment it is happy. Thus a hog is always happier than a man and a bacillus is happier than a hog" (p. 300 of Evans, 1968). The other half of the authors, Alexander Pope, for instance, take a wider view: [Happiness is] "our being's end and aim . . . that something . . . for which we bear to live or dare to die" (p. 299 of Evans). It would seem contradictory to die for the sake of a pleasant emotion. For Pope, happiness is life's bottom line. The same for Ovid: "None must be counted happy till his last funeral rites are paid" (p. 302 of Evans). And the same for Plato, as we saw in Chapter 3.

Aristotle's word for the ultimate aim of human life, *eudaemonia,* is usually translated as "happiness." J. L. Ackrill (1980) objects to this translation because he fears that people will understand the word in its emotional sense. Clearly that is neither Plato's nor Aristotle's sense. Here we will use the word *happiness* to refer to a good or harmonious pattern of life. Happiness thus understood "is best and better than everything else, not in the way that bacon is better than eggs and [better] than tomatoes (and therefore the best *of the three* to choose), but in the way that bacon, eggs, and tomatoes [are] a better breakfast than either bacon or eggs or tomatoes [alone]—and is indeed the best breakfast without qualification" (Ackrill, 1980, p. 21). The breakfast stands to its components, in Ackrill's analogy, as happiness stands to individual moral acts.

Aristotle says: "we have practically defined happiness as a sort of good life and good action" (*NE,* I, chap. 7, 1098b, 20). The happy person, according to Aristotle, needs at least some good luck ("blessedness") as well as moral behavior. A person who is repeatedly tortured cannot be happy no matter how moral her acts may be. A martyr, by virtue of moral actions, cannot be miserable. But she cannot be truly happy either.

Neither can a child. Since happiness is a molar pattern of human behavior, the person must have lived long enough for that pattern to have emerged and be exhibited. Prior to having lived a complete life, a person's happiness can only be provisional. In this, Aristotle agrees with Ovid, as quoted above, and with Yogi Berra: "It ain't over till it's over." But Aristotle puts it differently: "One swallow does not make a summer, nor does one day; and so too one day, or a short time, does not make a man blessed and happy" (*NE,* I, chap. 7, 1098a, 17). Here, Aristotle uses the word "make" as in "four lines do not make a pentagon," rather than as in "the salad-chef does not make dessert." Thus, for Aristotle, happiness, the aim of life, is a pattern of behavior. To see the aim of life as a specific internal state, cognitive or emotional, or even a specific act or a limited set of acts would be to compare a general class of things with a member of that class. It would be to make (in Gilbert Ryle's terms) a category mistake.

Moral Action

If happiness requires consistent moral action, how do we identify moral action: What actions are moral and what immoral? The defining characteristic of moral acts, according to Aristotle, is their location in the middle of a continuum. Bravery, as mentioned above, is a middle between rashness (fearlessness plus overconfidence) and cowardice (fearfulness plus lack of confidence). Temperance is a middle between self-indulgence and priggishness. Justice is a middle between too much for one person and too much for another.[7] Bravery, temperance, justice, and the like, however, are states of character. All people, according to Aristotle, are born with the capacity to be brave, temperate, and just. These characteristics are acquired not through teaching or instruction but by habit. A person who acts bravely must not only do brave acts but he must do them habitually. "It makes no small difference, then, whether we form habits of one kind or another from our very youth; it makes a very great difference, or rather *all* the difference" (*NE,* II, chap. 1, 1103b, 24; italics in original).

A few pages later: "Actions then are called just and temperate when they are such as the just and temperate man would do; but it is not the man who does these that is just and temperate but the man who does them *as* just and temperate men do them" (*NE,* II, chap. 4, 1105b, 5; italics in original). That is, if two people both perform the same apparently just act (say, they are storekeepers who return an overpayment to a customer), both people are not necessarily just and temperate. This act has to appear in the context of a series of other acts that form a pattern. The pattern may be labeled as just, but the individual acts are only provisionally just. An apparently just individual act might have been done merely to win praise or in the context of a promotional campaign or by compulsion or by accident.

Only a specific context—the other just acts—can make a given act just. This point may seem trivially obvious but there is a nonobvious distinction underlying it. For Aristotle a brave or just or temperate act is not a sign of an underlying internal state. If a just act were only a sign of a just character (as a high temperature is a sign of sickness), it would still be a just act regardless of its context. If the storekeeper had returned the overpayment merely to win praise or as part of a promotional campaign, the just act would be a sign of something that wasn't there—a false sign but no less just in itself. However, for Aristotle, the individual just act is part of a larger series of acts the whole of which constitutes a just character. If it is not part of such a series, it is not a just act. Analogously if the same series of three notes appears in two different overtures, one by Verdi and one by Puccini, say, an orchestra playing those three notes in the context of playing the Verdi overture is not in *any way* playing the Puccini overture. If you were to walk into the concert hall during those three notes and identify the overture as by Puccini, you would be wrong absolutely (however understandably). Your mistake was not about any hidden internal state but about perfectly *overt* external events, events that had taken place prior to your entering the concert hall and that would take place subsequently.

To put it another way, in judging from one brave act whether a person (you yourself or someone else) is brave, you are in the position of the blind man touching a real elephant, not of a sighted man looking at a fuzzy picture of an elephant. The evidence you have (the brave act) is part of the object you are judging (a brave pattern of behavior); the evidence is not a sign of an object that is somewhere else.

A person who is not habitually brave or just or temperate cannot become truly brave or just or temperate. If you do not naturally have these virtues you cannot acquire them. This is why Aristotle says their acquisition in childhood makes *all* the difference. However, that fact is not a disability but an advantage. A naturally virtuous person is virtuous by reason of appetites. Individual virtuous actions in themselves, are pleasurable to such a person. But given the variability of pleasure, natural virtue must be unstable. Pleasure is an attribute only of individual acts or a brief series of acts. If life is like building a house, then pleasure, as an object, might correspond to a good junction between a pair of bricks. It is highly unlikely that if your only object is to put pairs of bricks together, you will end up with a well-constructed house. That is why "strict virtue" requires reason. A person who cannot naturally put together a series of moral acts (who is not *temperate* by nature) may yet act like a temperate person by means of reason. (Aristotle calls such a person "continent" not "temperate.") We would say now that the person was *self-controlled*, or *mature*. The great advantage of strict virtue (continence) over natural virtue (temperance) is not only that consistent moral action constitutes happiness but *also* that the pleasure given up by going contrary to one's appetites (a pleasure anyway adulterated by the pain of deprivation required to establish those appetites) is more than compensated by the pleasure (a pure pleasure unmixed with pain) of reasoning when reasoning is used not to control appetite but to describe the abstract qualities of the universe (that is, to give scientific or philosophic wisdom).

Perhaps a biological analogy might clarify this point. Bees' wings were originally used not for flying but for cooling the hive (Heinrich, 1978). The use of an organ for a new purpose in evolution has been compared to the creative use of a human's reasoning capacity for purposes other than control of appetites. Aristotle was not a biological evolutionist but he did take this biological view of creative development. As noted previously, Plato saw the same facts from an ecological rather than biological perspective. He would have seen a flying bee as already present as a form in the ecological niche occupied by the nonflying (but hive-cooling) bee. The development of flight, then, would be merely the expansion of real bees further into this niche than they were before.[8]

Practical Wisdom

At the end of Chapter 2 the question was raised, What would be the nature of a behavioral account of the interaction of belief and desire? There is neither time nor space to provide such an account here. (Nor is an adequate nonbehavioral account available.) But we know the nature of a potential cognitive account. It would consist of a schematic diagram, much more complex but not fundamentally dissimilar to that at the right of Figure 2.2. Correspondingly, a behavioral account would have something to do with contingencies at the left of Figure 2.2. Although there are undoubtedly differences between *belief* and *reason,* on the one hand, and between *desire* and *appetite,* on the other, Aristotle offers a behavioral account of the interaction of reason and appetite that will make a good starting point when behaviorists come to account for interactions among abstract behavioral conceptions—that is, among mental events.

How, exactly, does reason interact with appetite? The modern cognitive/physiological outline of this interaction is clear. Reason would be located in the higher nervous system and would control appetite, located in the lower nervous system, much as a thermostat controls a furnace. The diagram in Figure 5.2b is a schematic outline of the conception. Particular instances would of course be much more complicated.

What is Aristotle's conception of the interaction of reason and appetite—practical wisdom? First, it bears repeating that for Aristotle the soul is in the body in the same sense that health is in the body (and not in the sense that the brain is in the body). The soul is in the body because the soul is the "moving principle" of the body. As T. N. Irwin (1980) puts it (p. 41), "The soul is the form because the form of the organism is its life— its goal-directed pattern of activity." Thus, when Aristotle distinguishes between voluntary and involuntary behavior by saying that when the moving principle is in the body, the movement is voluntary and when the moving principle is outside the body, the movement is involuntary, he is not referring to the presence of a mechanism (like the brain or the heart) physically located in the body and efficiently causing the movement. Rather, he is referring to the moving principle ("the goal-directed pattern of activity") into which this particular act fits: a final rather than an efficient cause. If the pattern of activity is characteristic of the person (is *in* the person), the act is voluntary. If the pattern is not characteristic of the person but of something else (is *outside of* the person), the act is involuntary. Thus, the very same act may be voluntary as seen in the light of one moving principle and involuntary as seen in the light of another.[9] A voluntary act in turn may be an instance of reason or appetite. The issue depends on how language may control behavior.

Language and the Control of Behavior

From the cognitive/physiological viewpoint the function of language is obvious. The experimenter's words enter the subject's nervous system (like the typist's strokes enter the word processor) and create an internal representation for further processing. The representation created is assumed to be the same as that created by an actual situation in everyday life. The experimenter's words are tied to the real-life situation by virtue of the common internal representation they both create. The internal representation is the *meaning* of both the words and the situation itself.

From the behavioral point of view the words are tied to the situation by virtue of the common overt behavior (rather than the common representation) they evoke. The function of the words is to place the subject's individual act into a specific pattern of actual environmental events. When an experimenter (studying human decision) says, "The probability of $100 is ½," the subject behaves as if a coin were being flipped and $100 were riding on heads or tails. The words are already a representation of the situation. No further representational mechanism is assumed to exist. Unless the subject has had some experience in the represented situation (with coins, wheels of fortune, dice, dredles, and so forth) the experimenter's words would be as meaningless to him or her as they would be to a pigeon. Language takes the "booming buzzing confusion" of events as they occur, occurred, and will occur, and picks out some pattern of those events as the context relevant to present behavior. In behavioral jargon, the experimenter's instruction is a "discriminative stimulus" (an S^D, in terms of Figure 2.2) for a complex behavioral pattern.

The sentences "Hot fudge sundaes taste good" and "Hot fudge sundaes make you fat" may both be true (in the sense that both are consistent with an individual's prior behavior), but each is a discriminative stimulus for a different act of choice. "Hot fudge sundaes taste good" sets the choice into a narrow context—the context of the *present state* of the eater's body. In the presence of a hot fudge sundae, the truth value of "Hot fudge sundaes taste good" is equivalent to "This hot fudge sundae will taste good now." On the other hand, "Hot fudge sundaes make you fat," is a general rule. It puts the present act of choice into a relatively wide context. But the sentence "Hot fudge sundaes make you fat" is *not* equivalent to its specific version "This hot fudge sundae will make you fat now." The more general statement is true but the specific version is false. You will not become fat now if you eat this particular hot fudge sundae.

The power of language to shift the context of particular acts is demonstrated in the previously described (p. 101) experiments by Mischel, Shoda, & Rodriguez (1989). Recall that in the standard version of these experiments, the children waited significantly longer (showed better self-control) with the pretzel covered than with the pretzel exposed to view. But these results could be reversed depending on what the experimenter told the children before she left the room. Children told to think of how crunchy, salty, and tasty the pretzel was, waited less time (were more impulsive), even with the pretzel covered, than children told to think of the pretzel as a log that might be used to build a toy house, even with the pretzel exposed.

Mischel's cognitive interpretation of these experiments relies on the influence of the child's thoughts ("hot" thoughts versus "cold" thoughts), caused by the experimenter's verbal statements, on the child's decision mechanism (as on the right of Figure 2.2). George Ainslie's (1992) cognitive interpretation assigns hot and cold thoughts not to the child but to internal "interests," within the child, that bargain with each other in an internal marketplace. The usefulness of such interpretations will depend on the procedures they suggest for manipulating choice and developing the child's ability to cope with real world problems. Neither of them are necessarily inconsistent with a potential behavioral explanation framed in terms of narrower versus wider temporal context. From a behavioral viewpoint the instructions to think of the pretzel's sensory qualities (like the words "hot fudge sundaes taste good") frame the present choice narrowly both spatially and temporally—within the child's body and now; the instructions to think of the pretzel as a log (like the words "hot fudge sundaes make you fat") frame the present choice widely—in the environment and in the future (when the toy house might be played with).

Theories of behavior concentrating wholly on cognitive control of emotions miss the point that the wider aspects of the environment (the toy log cabin) are valuable not just because they give rise to pleasant emotions or not just because they help us control and regulate unpleasant emotions but *in themselves*. "Subtility" of behavior (like subtility of a painting) lies not behind the behavior, not inside the organism, but in the patterns on the behavior's surface. A Mozart symphony is not better than a popular song because the emotions to which it gives rise are somehow better or "deeper" but because the pattern of the Mozart symphony is more complex and of longer duration; its context is wider, it is more abstract. And, as Plato pointed out, goodness lies in abstraction.

The most important function of language, therefore, is to expand the context of current behavior. Human happiness (perhaps the reader will agree) lies ultimately in molding one's life into an interesting pattern, one that is neither lost in its environment, nor one that clashes with it. Language, ranging from "pass the salt," to the contents of *Anna Karenina* is primarily a tool to achieve that end. Since our relations with other people constitute the most influential parts of our environments, I now go on to discuss the above considerations in the context of people's interactions in the social sphere.

Social Good

The opposition of particular and general rules, as it occurs purely with regard to individual pleasure and individual good, is the same as the opposition of particular and general rules as it occurs with regard to the individual good and the social good. Social welfare stands to individual welfare as individual welfare stands to individual appetite. In both cases, according to Aristotle, where conflict arises, the more abstract general principle should be the operating one.

For Aristotle, the social good outweighs the individual good, not because there is anything intrinsically good about altruism but because the basic principles of social welfare are generally more abstract than the basic principles of individual welfare. Since the facts derived from more abstract syllogisms are by definition more generally true than those derived from particular syllogisms, and since in cases of conflict a person's behavior should conform to the most abstract general principles that that person knows, Aristotle, like Plato, advocates the social over the individual good. For an individual, consistent failure to follow one's most abstract principles is, according to Aristotle, the *same as* (not just evidence of) a failure to know those principles. A person may fail occasionally but consistent behavioral failure *means* lack of knowledge. Verbal articulation of a principle is indeed evidence of knowledge of that principle, but consistent ethical behavior is the knowledge itself. To repeat, verbal behavior without ethical behavior, according to Aristotle, "means no more than its utterance by actors on the stage."

The political philosopher will therefore try (as Plato did in *The Republic*) to design a society in which the rewards and punishments are such that ethical behavior (from the point of view of the philosopher) is identical with pleasurable behavior (from the point of view of the individual). Obedience to the laws of such a society is, as Aristotle says, the highest form of virtue. But for Aristotle, unlike for Plato, a perfect society needs something more: a mechanism whereby each individual obeys the law (is good) not because of the pleasure inherent in such acts of obedience (as would be the case in a perfect society) but because of the (purer) pleasure inherent in the understanding of the law. To pursue the music analogy further, an opera may be enjoyed throughout because it has good local action and melodious local themes. But the opera would be still more enjoyable (with pure enjoyment) if, in addition, its overall structure were understood.[10] Aristotle resolves the tension between individual virtue and social intelligence in a perfect society by adding the desirability of philosopher-citizens to that of a philosopher-king. In imperfect (that is, actual) societies, as previously indicated, Aristotle and Plato agree that individual excellence (goodness for Plato; knowledge for Aristotle) can be achieved only through philosophy.

The perception of mental entities in behavior, even the most complex human behavior, is the start of, not the end of, psychological investigation. As behavior becomes still more complex, as desires and beliefs become more abstract, as we see them more and more exclusively in humans, there is nothing at all to close off one line of empirical investigation or another. Some economists (Becker, 1976, for example) have attempted to account for complex individual human behavior with utility functions such as those of Figure 3.2. Teleological behaviorists need be no less presumptuous.

Chapter 6 attempts to provide a plausible historical (and almost wholly speculative) reason why we believe at present that our mental lives occur inside us (as water is inside a jug). Although the chapter focuses on St. Augustine and Descartes, there is no attempt to trace present thought to these philosophers' direct or even indirect influence. It may be supposed, rather, that the forces that shaped their beliefs might also have given rise to that part of our folk psychology that has developed into cognitive science. Once we articulate these forces, we can ask (as we do in Chapter 7) whether they need to have absolute dominion over modern psychology.

Notes

1. It is on this point that the present behavioral reconstruction differs most crucially from most modern interpretations of Aristotle. As Chapter 2 indicates, modern Aristotelians like Charles Taylor (1964) and Alan Donagan (1987) more or less strongly oppose behaviorism.

2. This is a different question from the modern one of what it's *like* to be something. Thomas Nagel's (1974) essay "What is it like to be a bat?" tries to draw a line between conscious and unconscious beings. If it is *like* something to be a bat, then a bat is conscious. The problem with this sort of endeavor is that the attribution of consciousness to an animal is in a sense honorific. At one recent seminar an eminent philosopher and an eminent biologist debated the question of whether bats were conscious. The philosopher denied consciousness to bats on the grounds of their inability to treat other animals as if *they* were conscious. However, for each citation by the philosopher of bats' failure to pass certain critical tests, the biologist, who had spent his life and made his reputation studying the behavior of bats, cited an equivalent success. Apparently, the better you know bats, the cleverer they seem within the context of the environment they face. Since the biologist knew bats better than anyone else at the seminar, he won that particular argument. The feeling of at least one observer of this debate was the same one he had in the movie when the Wizard of Oz bestows a diploma on the scarecrow.

In the nineteenth century a similar and equally pointless debate took place between the philosopher Rudolf Hermann Lotze (1817–1881) and the biologist Edward F. W. Pflüger (Boring, 1950, p. 29). The biologist argued that consciousness exists in the spinal cords of frogs because a frog is capable of purposive movement after its head is cut off (purposive movement at that time being the agreed criterion of consciousness). The philosopher argued that the particular purposive movement in question (withdrawal of a leg to avoid a painful stimulus) could be *learned* by the frog only with its head on and thus did not in itself indicate consciousness (thus slightly altering the criterion). Pavlov's later demonstrations of the reflexiveness of apparently purposive movement blunted the point of the issue.

3. Swerving in this situation may also be conceived as something more than just a discriminative act. To consider only its discriminative function is to consider the act as a perception. To consider its function in preserving the life of the driver is (as we shall see in a moment) to consider the act as a thought.

4. Religious interpreters of Aristotle such as St. Thomas Aquinas (1225–1274) and, in more recent times, Franz Brentano (1838–1917), have naturally

shied away from his extreme behaviorism. Concepts that make perfect sense when interpreted in terms of overt movements of whole organisms and the consequences of those movements are obscured by the introduction of an incorporeal soul which may move while the body stands still or by "intentions," which (perhaps distorting Brentano's own intentions) are now frequently supposed to efficiently cause the body's movements.

5. Just as the basic principles of geometry (its *archae*) are outside of geometry itself, so the basic principles of philosophy must be outside philosophy. Aristotle therefore has little to say explicitly about the rationale for using reason to understand the abstract nature of the world (except to praise the philosophical life in the last of the ten chapters of the *Nicomachean Ethics*). However, the Aristotelian corpus as a whole may be thought of as an ostensive argument for such a life.

6. Of course, being human, we are faced with a problem of allocation of our limited reasoning ability (in terms of Figure 5.2, limited R'). Some must of necessity go to regulating our appetites (S*). Otherwise, they would dominate our lives (R* would be the same as S*). The remainder, however, may be used as "pure reason," an attribute of which is pure pleasure. In economic terms the problem is not unlike the allocation of a country's resources to defense versus consumption. Only the gods do not have this allocation problem. According to Aristotle, the gods devote their time wholly to reason.

7. Aristotle conceives of two kinds of justice as characteristic of individual behavior: first, in the relation of individuals to the state where justice is equated with obedience to the law. If the laws are good, obedience to the law is the highest form of excellence—the highest virtue. "Justice in this [first] sense, then, is not [merely] part of virtue but virtue entire" (*NE,* V, chap. 1, 1130, 10). In the second and more particular sense, justice is conceived as a largely economic relationship between individuals. Just acts are redistributions of honor, money, and even pain (being stabbed is a "loss" that may be compensated by stabbing conceived as a "gain") to restore a balance between individuals. A judge is a mediator (a middle-person might be a better term) whose job is to restore a balance (a middle) that has been upset.

8. According to Plato's conception, real bees could never be ideal bees for two reasons: first, because they could never exactly fit into the existing niche but, second, and more important, the existing niche itself is not ideal because it is formed out of the imperfect environment of the bee. The pure Form of a bee would require, for instance, perfect flowers. Similarly, for Plato, ideal human behavior could only occur in an ideal society.

9. One example that Aristotle uses to illustrate this point is "the [captain's] throwing of goods overboard in a storm; for in the abstract no one throws goods away voluntarily, but on condition of its securing the safety of himself and his crew any sensible man does so" (*NE,* III, chap. 1, 1110a, 9). From the point of view of a less abstract principle (maintain the safety of the ship), the act is voluntary (and the captain is responsible for it) because the principle is *in* the person who does the act (in the sense that it is consistent with the captain's general pattern of behavior), while from the point of view of a more abstract principle (throw away the owner's worldly goods), he act is involuntary (and the captain is not responsible) because the principle is *outside* the person who does the act (in the sense that it is inconsistent with the captain's general pattern of behavior). The fundamental distinction between voluntary and involuntary behavior thus does not lie in different internal

states but in different external perceptions. It is not, What is the way things are? but instead, How ought we to view them? How are they related to the social system?

10. A better analogy than watching and hearing an opera might be *performing* an opera. Recall that the conflict between appetite and reason is basically a conflict of perceptions (categorization of particular facts) but pleasure, when it occurs, is not a perception at all. Aristotle says: "It is not right to say that pleasure is a perceptible process, but it should be called activity of the natural state and instead of 'perceptible' 'unimpeded.'" It is important to note, however, that for Aristotle the attribute of an unimpeded natural action that is pleasurable is not the movement involved (movement is an attribute of action but pleasure is not movement) but the naturalness and unimpededness of the action. For Aristotle, unimpeded natural actions are inherently pleasurable by definition. No state (habitual pattern of actions) can be impeded by the pleasures naturally arising from it (by any unimpeded local actions that may comprise it). However, a habit could be impeded by an incidental pleasure. For instance, a person who loves music is hindered in arguing a point if she hears a flute playing because the activity of listening to the flute interferes with the activity of arguing.

6

Augustine and Descartes: The Concept of Free Will

About Augustine, J. R. Kantor has this to say (1963, p. 280): "Augustinian philosophy represented a blind acceptance of mystic postulates fashioned to achieve personal salvation and a transcendental refuge from the obvious evils of the day. What can be the method of such a philosophy? Nothing other than verbal dialectic. Having accepted the popular assumption that there is a Christian truth and that there can be no other, Augustine knows the answer to all questions and problems in advance."

"Descartes," Kantor (1969, p. 50) says, "stands firmly in the spiritualistic tradition and thus some of his focal doctrines go straight back through St. Thomas to St. Augustine." Why then have both Augustine and Descartes had such a tremendous influence on human thought up to the present day? This chapter explores some of the weaknesses of Aristotelian behaviorism that made it inadequate to guide people's lives (including their thoughts) in the worlds in which Augustine and Descartes lived.

Augustine

While Augustine was certainly not a behaviorist philosopher, he was nevertheless confronted with objective problems that the philosophies of Plato and Aristotle were apparently unable to solve.[1]

The ancient Greek society in which Plato and Aristotle lived, although diverse and full of conflict, was nevertheless fundamentally orderly. The Greek city-states were united by common language, customs, and religion and by a common enemy: Persia. Although their governments differed, both Plato and Aristotle were able to find common features among them. The perceptibility of social order perhaps enabled the ancient Greek philosophers to have faith in the order of all natural things. The central Greek problem, the relationship of the one to the many—the abstract to the particular—was therefore approachable by both Plato and Aristotle as a problem in and of nature. For both Plato and Aristotle, abstract and particular were both ways of classifying the world. Although Plato and Aristotle differed in where they drew the line between perception and cognition and in what they were willing to call "reality," both philosophers viewed perception, cognition, and all of mental life as modes of interaction of a person with the world.

The late Roman world in which Augustine lived and traveled was much more diverse than the Greek world. Each part of the empire had different customs, different local religions, different languages—much like the modern world. But whereas we have adopted a common physics, biology, and technology that overarches social disorder, there was no such set of natural principles to which Augustine could refer. Two fixed points in Augustine's world were (1) the ability to reason, common to all people, and (2) the word of God.

Nested within the problem of the relationship of abstract and particular is the ethical problem, not really solved by Plato and Aristotle, of the relationship between society and the individual. The ethics of both Plato and Aristotle are elitist. For both Plato and Aristotle, a good life is achievable, in an imperfect world, only through individual intelligence. For Plato, a perfect society would enable everyone to live a good life because such a society would be arranged so that pursuit of pleasure would coincide with goodness. But in an imperfect society—a real society—only a philosopher can discriminate accurately between goodness and pleasure. For Aristotle, the good life is a delicate balance (a mean) between opposites, achievable only by constant intelligent vigilance. The behavior of the rest of humanity—people who cannot discriminate so finely—is subject, according to the ancient Greek philosophers, to the vagaries of the laws of imperfect states.

Augustine dealt with the metaphysical problem of the relationship between abstract and particular by separating the former from the latter, and locating the abstract entirely and exclusively where the most order was to be found: in human reason. He dealt with the ethical problem by making access to these now internal abstract principles depend not on individual intelligence but on individual willpower, which any person is free to exert. By these conceptual moves Augustine was able to incorporate religion into philosophy much more harmoniously than could Plato or Aristotle who, while they both held piety to be a virtue, nevertheless kept the gods in heaven, conceptually separate from human society.

Much of Augustine's work is in fact strongly religious in character; its object is not to convince but to guide the behavior of believers; his earlier writings, however, are more philosophical than his later ones. J. H. S. Burleigh (1953, p. 107) regards the long essay "On Free Will" (FW) as "the high water mark of [Augustine's] earlier works, and the best and fullest exposition of what may be called the peculiarly Augustinian brand of Neoplatonism." This work, a dialogue between Augustine and a follower, Evodius, is therefore worth discussing.

"On Free Will" is part of an argument against Manichaeanism, the belief previously held by Augustine himself that good and evil are equal and opposite forces in the world, that there exists "an evil nature, unchangeable and co-eternal with God" (*Retractions* I, ix, 1). In "On Free Will" Augustine argues that evil is a consequence of human free will. God's highest creation is a person who is good of his or her own free will. Such a person is better than one who is good by compulsion. But people who are free to be good must also be free to be bad.

Augustine's conception of free will implies a view of causation different from Aristotle's. According to Aristotle, all movement has both an efficient and a final cause. Freedom for Aristotle is a category of behavior, the final cause of which is expressible in terms of purposes of the moving object. Thus, for Aristotle, even a stone may be free when its behavior is explicable in terms of its own habit (that is, its own object)—to head toward the center of the earth. If it had any other object (which it would have to have if it moved upward), its behavior would have to be explained in terms of other purposes (say, of the person who threw it) and to that extent the stone would not have been free.

For Augustine, free will is a break in the normal chain of efficient causation. Human voluntary behavior does not have a chain of efficient causes traceable to God in the sense that a stone does. God is the ultimate efficient cause of all motion *except* voluntary human behavior. The chain of efficient causation of voluntary human behavior is traceable backward only as far as the free will of the behaving person.

Human motion, which for Aristotle is behavior of a whole human organism in the world, is interiorized by Augustine, who conceives of a metaphorical "motion of the soul" upward toward virtue or downward toward sin: "It is not in the power of a stone to arrest its downward motion, while if the soul is not willing it cannot be moved to abandon what is higher and to love what is lower" (FW, III, i, 2).

On the other hand, all voluntary human behavior does, according to Augustine, have a *final* cause. Human final causes are divisible into two main categories: virtuous and sinful. Virtuous final causes are identified with God; sinful final causes are identified with the pleasures and pains of worldly objects. The critical question then is, How does a person aim his free will "upward" so as to achieve goodness and avoid sin? It is in answer to this question that Augustine makes his distinctive philosophical contribution.

Like Plato, Augustine identifies wisdom with the highest virtue (he uses Wisdom as another name for Christ). For Augustine, however, as opposed to Plato, wisdom and virtue are separable from the world not only conceptually, not only in thought, but also in essence. Both Plato and Augustine use visual analogies to make the distinction clear. In Plato's allegory of the cave, the wise man sees the world as it really is; he sees the world differently from the way the chained prisoners see it, but still it is the world he sees. The prisoners see merely a projection of the world, a two-dimensional (particular) version of it. The wise man sees that same world in three dimensions (abstractly). For Plato, therefore, wisdom is analogous to a better view of something but not a view of a different thing. Plato's wise man could not gain wisdom by closing his eyes; on the contrary, his eyes take in more than those of the prisoners. The wise man's vision might better be called *outsight* than *insight*.

But Augustine separates wisdom and virtue entirely from the world. He identifies them, in substance, with God and locates them inside the person. So the visual analogy is not to seeing better or seeing more dimensions of some thing, but to seeing something else entirely—to looking in a different direction—away from the world and into oneself. Thus, for Augustine, the analogy to upward movement is interior vision. To put it another way, for Plato and Aristotle, the wise man directly perceives the abstract nature of the world (which for Plato is the only *real* world); for Augustine, the wise man perceives his own reason: "Does reason comprehend reason by any other means than by reason itself? Would you know that you possess reason otherwise than by reason? . . . Surely not" (FW, II, 9).

As previously indicated, for Plato, reflection may be interpreted as external reflection, taking place in the world—from the world to the person and back to the world, an "external feedback loop" in modern terms. A painter painting a scene would be reflecting the scene. For Augustine, the feedback loop is wholly internal. This concept of internal reflection or introspection was not invented by Augustine; it seems to have been imported by him into Western philosophy via Plotinus from Eastern mysticism.[2] It is of course with us now, deeply imbedded in our language, especially the language of mental terms.[3] However, an important difference between introspection as conceived by Augustine and the modern version is that, for Augustine, despite the wholly internal nature of reflection, what we see when we reflect is *not private*. Just as, when we look to the outside world we see something common, so do we see something common when we reflect inwardly: God.

It is tempting for the nonreligious reader to dismiss Augustine, who certainly argues for the primacy of faith over reason. But, in his early writings at least, Augustine treats the concept of God analogously with Plato's conception of the Good. In Augustine's view God stands for all abstract truths—all known rules—ranging from mathematical rules ("the science of numbers") to social rules. The rule that seven plus three equals ten and the rule that "each man should be given his due" are alike absolutely and eternally true: they are therefore identified as "wisdom" (FW, II, x, 28) and are not empirical.

Augustine agrees with Plato that abstract truths are not derivable from observations of the behavior of particular objects in the world. Neither, certainly, are they mere social, linguistic, or logical conventions. According to Augustine, they are eternal truths but observable in only one way: by internal observation (or reflection or introspection). The turning of vision from observation of the external to the internal world is the main function of our free will.

Because the concept of introspective knowledge is so familiar to us, it is important to emphasize again how Augustine's conception of introspection differs from ours. In the modern conception, introspection gives us *private* knowledge, available to ourselves alone. For Augustine, however (in this early essay on free will), what we see when we turn our vision inward is not private and subjective at all, but *public* and *objective,* more truly objective—because unchangeable and the same from all perspectives— than what we see when we look outward.

Augustine considers whether mathematical truth might be empirical and rejects it:

AUGUSTINE: Suppose someone said that numbers make their impression on our minds not in their own right but rather as images of visible things, springing from our contacts by bodily sense with corporeal objects, what would you reply? Would you agree?

EVODIUS: I could never agree to that. Even if I did perceive numbers with the bodily senses I could not in the same way perceive their divisions and relations. By referring to these *mental operations* [italics mine] I show anyone to be wrong in his counting who gives a wrong answer when he adds or subtracts. Moreover, all that I contact with a bodily sense, such as this sky and this earth and whatever I perceive to be in them, I do not know how long it will last. But seven and three make ten not only now but always. . . . So I maintain that the unchanging science of numbers is common to me and to every reasoning being.

AUGUSTINE: I do not deny that your reply is certainly most true.

        (FW, II, viii, 21, 22)

Figure 6.1 illustrates Augustine's conception of the two sorts of vision—external and internal. Particular objects (created by God) are directly sensed and controlled by individual people. Augustine's internal sense corresponds to Aristotle's "common sense," which nonhuman animals may possess. But whereas Aristotle's common sense is interpretableas a common subpattern in two overt patterns of discrimination (see

EXTERNAL WORLD   **SOCIETY**   **INTERNAL WORLD**

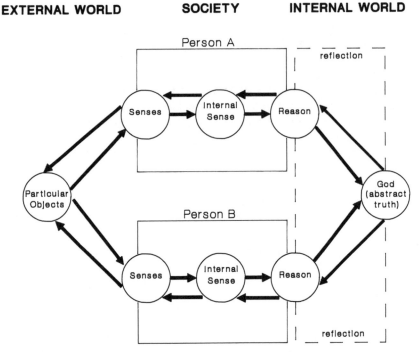

Figure 6.1 Augustine's conception of the interaction of a person with the
world, on one hand, and with God, on the other. The right-ward arrows
represent direct knowledge. The left-ward arrows represent judgment and
control. The loops in the dashed box represent internal reflection.

Chapter 5), Augustine's internal sense has the status of what we would
today call a "cognitive mechanism" or an "information-processing mod-
ule" that stands in humans between particular senses and reason. The next
step in the process, the loop between reason and God (by which reason
knows itself) possessed by humans, is Augustine's version of reflection or
introspection. This loop, like the loop between a person and the external
world, contains an object common to anyone who observes it. Whether
people orient their vision one way or the other (to the left or right in
Figure 6.1) is voluntary but, either way, the view is common. Just as "there
is one object for both of us, and both of us see it simultaneously" (FW, II,
vii, 16), so also "these rules and guiding lights of the virtues . . . are true
and unchangeable, and singly or all together they stand open for the
common contemplation of those who have the capacity to behold them,
each with his own mind and reason" (FW, II, x, 29). The difference, again,
is not that one orientation is to public and the other to private objects
(both directions lead to public objects), but that one orientation is to
temporal and the other to eternal objects.

Plato describes a virtuous life as reason in control of ambition and
appetite. For good self-control, reason, like a charioteer, has to see further

than his horses. Still, the horses are necessary to pull the chariot. For Aristotle, the (fundamental) nutritive and sensible souls are an essential part of the rational soul (like the three points of a triangle are part of the four points of a square). Exercise of the higher soul, for both Greek philosophers, is thus *through* the lower soul as it acts in the world. But for Augustine, regardless of how intensely you focus on the world, regardless of how well your reason controls your appetites, you will not reach your highest individual (hence the social) good. For Augustine, appetite must not be just controlled; it must be suppressed entirely: "The lover of the good will resists and opposes lusts in every way, and so is rightly called temperate" (FW, I, xiii, 27).

Plato and Aristotle both admit that people may be intelligent and yet bad, because they may use their intelligence to gratify rather than to control their appetites. For Augustine, degree of intelligence does not matter at all with regard to the highest individual good. Instead, what is required is the *will* to turn one's gaze inward instead of outward. Once you do that, Augustine says, God will appear to you as vividly and as automatically as a mountain does when you look at it. The critical factor in this process of turning inward is not good reason but good will. Reason is necessary in this process, as eyes are necessary to see. But even people with weak eyes can see a mountain and even people of weak intelligence can be good, according to Augustine, in the highest sense. For those who turn the other way (to the left in Figure 6.1), Augustine would agree with Plato that "the penalty they pay is the life they lead," a life, in Augustine's terms, "which is better called death" (FW, I, iv, 10).

It may be argued (as Kantor probably would have) that Augustine had a political motive for deemphasizing the role of intelligence in ethics. Augustine had not yet become a bishop when he wrote "On Free Will." But he was nevertheless a devout Christian who wished to convert others to the faith. A religion, still with many competitors, could not be successful if it rejected potential adherents (as Plato's and Aristotle's philosophies undoubtedly did) on the grounds of intelligence. But even if political, it is important to note that Augustine was offering more than happiness after death to adherents of his philosophy. Like Plato and Aristotle, he was attempting to discover a mode of happiness during one's lifetime on earth. Near the beginning of "On Free Will," he states his criterion: "To desire to live without fear is characteristic of all men, not only of the good but also of the bad. But there is this difference. The good seek it by diverting their love from things which cannot be had without the risk of losing them. The bad are anxious to enjoy these things with security and try to remove hindrances so as to live a wicked and criminal life which is better called death" (FW, I, iv, 10).

Although intelligence itself cannot be taken away, people who use their intelligence to increase their pleasure (bad people or people with bad will) will be unhappy because what they use intelligence for *can* be taken away. Augustine threatens the bad person, not with hellfire only, but also with unhappiness on earth. Correspondingly, the good person will be happy because that person has something that cannot be taken away—not just in practice but in principle.[4] Augustine is eloquent on this issue:

> The beauty of truth and wisdom [identified by Augustine with God], so long as there is a persevering will to enjoy it . . . does not pass with time or change with locality. It is not interrupted by night or shut off by shadow, and is not subject to the bodily senses. To all who turn to it from the whole world, and love it, it is close at hand, everlasting, bound to no particular spot, never deficient. Externally it suggests, internally it teaches. All who behold it, it changes for the better, and by none is it changed for the worse. No one judges it, and no one without it judges aright. Hence it is evident beyond a doubt that wisdom is better than our minds [therefore identifiable with God], for by it alone they are made individually wise, and are made judges, not of it, but by it of all other things whatever. (FW, II, xii, 38)

The reward for turning to wisdom "from the whole world" is happiness:

> What do you ask for more than to be happy? And what is more happy than to enjoy unshakable, unchangeable truth which is excellent above all things? . . . Men exclaim that they are happy when with throats parched with heat they find a fountain flowing with pure water, or being hungry, find a copious meal all ready prepared, and shall we deny that we are happy when truth is our meat and drink? (FW, II, xiii, 35)

Thus, true happiness, the highest individual goodness, and the social good are drawn away by Augustine from the intelligent few and placed in the hands of any human being. In modern terms, the individual self is seen by Augustine as a motivational self rather than a cognitive self.

The notion that happiness is available only to an elite few is of course not eliminated by this transformation. Augustine argues at length that God retains foreknowledge of who will exercise free will for good, and who for evil. Augustine has merely substituted a motivational elite for an intellectual elite. A member of Plato's intellectual elite might be good or evil (then as now intelligence does not guarantee virtue) but for Plato, in order to be truly virtuous (in a real society), a person had to be intelligent. On the other hand, a member of Augustine's motivational elite was automatically virtuous—in the highest sense.

Augustine's Neoplatonism

Excellence of the self (self-control, human happiness) is seen by Augustine in motivational rather than cognitive terms. But the modern-day behaviorist makes no distinction between motivation and cognition *as* internal processes. What then, from the point of view of modern behaviorism, is the essential difference between Plato and the Neoplatonist, Augustine? Imagine Figure 6.1 rolled about the vertical axis so that the circle labeled "God" overlaps the circle labeled "particular objects" and the area labeled "internal world" overlaps the area labeled "external world." Plato's philosophy could then be considered as a sort of rolled-up version of Augustine's philosophy, and Augustine an unrolled version of Plato. In the flat, unrolled Augustinian diagram, "free will" determines in which direction the self will turn. Human effort to achieve happiness is a matter of internal orientation. The abstract has *already* been separated from the particular so there is no cognitive work to be done. Once a person's motivational work is over, control of her individual overt behavior is in the hands of God.

In the rolled-up Platonic diagram, the abstract overlaps the particular and must be teased apart from it by reason. Once this cognitive work is done there is no motivational work left to do. To *see* something clearly better and something clearly worse is to without further effort choose the better. Thus, in behavioral terms there is no difference between Plato's exercise of the intellect and Augustine's effort of the will: both types of effort result in behavior under the control of abstract entities.

In behavioral terms, the difference between Plato and Augustine must be, not in the internal processes by which good behavior is achieved but in the objects of that behavior: the value systems or the "final causes" toward which it is aimed. What, then, is the difference between Plato's Good (which he sometimes calls God) and Augustine's God (which he sometimes calls Good)?

For Plato, good individual behavior meant behavior under the control of the environment conceived in its most abstract (hence longest-lasting) sense. Other human beings were seen by Plato merely as part of that environment. The wise person's behavior differs from the foolish person's behavior, according to Plato, by fitting better with nature (including human nature).

Augustine, by (metaphorically) unrolling the diagram of Figure 6.1, separated abstract truth from nature as a whole but kept it *common to all human beings*. Thus Plato's wider abstraction from nature is narrowed by Augustine to an abstraction from humanity. For Plato, the ultimate truth is *natural;* for Augustine, the ultimate truth is *psychological* and *social*. The essential difference between Plato and Augustine then comes down to the importance they attach to human beings relative to the rest of the universe. Contrary to Kantor, one philosopher is no more transcendental than the other. Both may be thought of as natural in the sense that both found truth and goodness in abstract conceptions of the environment. Augustine's conception of the relevant part of the environment was simply narrower than Plato's.

Augustine and Behaviorism

Emphasizing again that Augustine was far from behaviorism and recognizing that many antibehaviorist attitudes originate in various interpretations of Augustine, it is nevertheless instructive to analyze Augustine's philosophy from a behavioral perspective.

Introspection, in opposition to which behaviorism originally rose, loses its principal disadvantage as a psychological method when its object is not intrinsically private. As conceived by Augustine, introspection was a mode of focusing on the most abstract conception of individual and social behavior. Internal reflection could not be in error, according to Augustine, for the same logical (as opposed to empirical) reason as that by which sensation could not be in error according to Aristotle: the concept of internal reflection already implies success; it implies seeing rather than looking. The motivational effort and ability involved in reflection is no different from the cognitive effort and ability involved in abstraction: both are operationally identifiable as living a good life, a life harmonious in the long run with nature (for Plato) or with society (for Augustine).

Remember that Plato's ideal world is not a perceptual one but a functional one (the ideal chair is a functional, not a perceptual, concept). Similarly, Augustine's ideal society (the "City of God") is a functional rather than a perceptual one; it has nothing to do with real societies (the "City of Man"). The ideal society is public but not (externally) perceptible. Introspection is thus, for Augustine, something like *insight* for the gestalt psychologists, a direct awareness of the functional relationships of things—not a private internal vision with the same properties as external vision.

The "internal feedback loop" that constitutes Augustinian internal reflection has two focuses: reason and God. Conceptually, they are separable; reason is the center of psychological truth and God is the center of social truth. But the action of internal reflection is a single process. Thus, an important modern question, Is individual rational behavior consistent with the social good? is answered by Augustine in the affirmative. Individual reason, when oriented correctly by the will, necessarily encompasses the social good. The knowledge that seven plus three equals ten and the knowledge that man ought to live justly are for Augustine both the same sort of knowledge. The Bible was, for Augustine, a book with the status of a mathematics text—one in which eternal laws have been written down.

Faith, for Augustine, is thus a generalization and expansion of what Skinner (1957) calls "rule-governed behavior" (see also, Hayes, 1989). When a parent tells a child "You will be happier if you study than if you go to the moves tonight," the parent assumes that the child has a certain faith. The child gives up present pleasure, which it knows from past experience is contingent on moviegoing, for a future reward, which it does not know from past experience is contingent on studying. The rule the child follows is: Obey your parents. Augustine would agree with Aristotle that happiness cannot be achieved in a brief time. Thus, *while* it is being achieved one has to have faith that the long behavioral path to it is the correct one. Eventually the child might learn that studying is more rewarding than moviegoing; then the child knows that in which it previously had faith.

Belief in this sense precedes knowledge. Augustine says, "we must first believe before we seek to know" (FW, II, ii, 5) and quotes the Bible in support: "'Except ye believe ye shall not understand' (Isa. 7:9, LXX)."

As rules become more abstract, it takes more and more experience to gain knowledge. "Obey your parents," for instance, is a more general rule that takes longer to obey and that requires correspondingly more faith than does "study tonight." The most general rule, "obey God," requires the most faith. But just as a child can eventually know that "obey your parents" is a good and true rule, a person can, according to Augustine, eventually know (without having to die first) that "obey God" is a good and true rule.

Descartes

For Kantor, Descartes (together with Hobbes) occupies a position two steps up (from transcendentalism toward naturalism) from St. Augustine.[5] The first step is occupied by St. Thomas Aquinas, mainly because of his interest in promoting Aristotle's *De Anima*. But Aquinas retained Augustine's unfolded version of Figure 6.1. Kantor (1963, pp. 345–346) says:

> Once the ship of internality became launched in the intellectual sea, a whole host of problems followed in its wake. . . . How can the internal be matched with outside things? . . . Aristotle simply treated the problem in terms of the organism and the object with which it was in interaction. But the Thomist had to add his spiritual factors. . . . All the changes in the basic scholastic doctrines developed by the succession of mentalistic psychologists, for example, the making of the external object into an [internal] stimulus or the addition and interpretation of neural connection and brain termination, have simply kept viable the inner-outer tradition.

Aristotle's philosophy lent itself well to religious interpretation. Aristotle's conception of final causes as epistemologically prior to efficient causes gave religious leaders, who claimed to have direct access to final causes (by means other than those conceived by Aristotle), reason to hold their knowledge superior to any kind of empirical observation (regardless of how penetrating). Thomistic philosophy may be seen as an interpretation of this sort.

Between Aquinas and Descartes, however, the rise of Renaissance natural philosophy (via Copernicus, Kepler, and Galileo—"scientists," in modern terms) made claims of superiority on these grounds more and more suspect. The physical world came to be understandable in terms of efficient causes alone. Final-cause explanations of human action could thus be seen as less valid, scientifically, than efficient-cause explanations of human action, which were just then being promoted. William Harvey's (1578–1657) discovery of the circulation of the blood and the function of the heart as a mechanical pump was only the most successful example of a widespread movement toward mechanism among Renaissance physicians and anatomists. Descartes, himself a major contributor, faced the task of reconciling this movement with a religious conception of the human soul. He did this by rejecting final-cause explanations altogether in the realm of human behavior as well as the behavior of nonhuman animals and of physical objects. All human behavior, according to Descartes, was efficiently caused; furthermore, the ultimate efficient cause of all human behavior was God. Renaissance science was thus combined with religion.

One class of human behavior—automatic, relatively simple motions, such as sneezing, pulling one's hand from a fire, focusing one's eyes—was explained by Descartes in terms of chains of efficient causes (later called "reflexes") originating in the environment (and ultimately in God as the creator of the world). This class of human behavior could be studied, according to Descartes, in the same way as Galileo had studied the behavior of physical objects: by observation and experiment.

This was a major concession to Renaissance science; it put all of physiology and biology as well as physics outside the authority of the church. The natural philosophers who studied the world (including physics, biology, and physiology) were, by this concession, acknowledged not only to understand the world better than religious authorities could, they were also acknowledged to be gaining a better (if still indirect) understanding than religious authorities could of God as the creator of the world. It is not surprising, therefore, that Descartes withheld publication of these speculations during his lifetime.

The one exception to this Cartesian concession to empirical science, however, was psychology. One class of human behavior only—purposive movement such as speech, social behavior, and learned behavior—was exempted by Descartes from experimental study. This class of behavior was no less efficiently caused than the rest; purposive behavior was said to follow the same neural and muscular paths in a human being as reflexive behavior. And, just as reflexive behavior was held to be ultimately caused by God, so was purposive behavior. But the aspect of God that Descartes conceived as the originator of purposive human behavior was not represented in the mechanical world (and knowable indirectly by empirical observation); it was the directly knowable (Augustinian) aspect of God represented in the souls of individual human beings.[6]

All human behavior was thereby brought by Descartes under the umbrella of efficient causation. The ultimate cause of voluntary human behavior was said to be inside the behaving person, directly knowable by that person but not empirically knowable by another person, because external efficient causes of voluntary behavior were held not to exist.

This division by Descartes exempted psychology from empirical study on two grounds: (1) Empirical study of the human soul is impossible because there are no external, observable efficient causes of purposive behavior; and (2) Even if such an empirical study were possible (say, by observing voluntary behavior and inferring its internal cause), it would be pointless because there is a still better way to understand the human soul—by internal reflection (introspection).

Aristotle's conception of psychology as a branch of physics was thus deliberately abandoned by Descartes, together with Aristotle's conception of a science of final causes. The history of psychology since Descartes may be seen as a series of attempts to somehow make psychology an empirical science—to achieve in psychology the success of physics and physiology either by applying their observational techniques to introspection or by reducing psychology to one or the other, without at the same time accepting final-cause explanations (teleological explanations) of psychological phenomena. It is therefore worth examining Descartes's philosophy further.

Descartes's Method

The method of using one's own doubt to derive certainty was adopted by Descartes from Augustine. The following passage is from Augustine's *On The Trinity (De Trinitate*, x, chap. 10, 13–16; quoted by Kantor, 1963, vol. 1, p. 187):

> Who ever doubts that he himself lives, and remembers, and understands, and wills, and thinks, and knows, and judges? Seeing that even if he doubts, he lives; if he doubts he remembers why he doubts; if he doubts he understands that he doubts; if he doubts he wishes to be certain; if he doubts he thinks; if he doubts he knows that he does not know; if he doubts he judges that he ought not to assent rashly. Whosoever therefore doubts about anything else, ought not to doubt about all these things; which if they were not, he would not be able to doubt of anything.

Here is a passage from Descartes's *Meditations* (1641 / 1971, edited and translated by E. Anscombe & P. T. Geach [A & G]):

> Am *I* not the very person who is now "doubting" almost everything; who "understands" something and "asserts" this one thing to be true, and "denies" other things; who "is willing" to know more, and "is unwilling" to be deceived; who "imagines" many things, even involuntarily, and perceives many things coming as it were from the "senses"? Even if I am all the while asleep; even if my creator does all he can to deceive me; how can any of these things be less of a fact than my existence? Is any of these something distinct from my consciousness? Can any of them be called a separate thing from myself? It is so clear that it is I who doubt, understand, will, that I cannot think how to explain it more clearly. (Second Meditation, p. 70)

The difference between Augustine and Descartes is that, when Augustine looked inward and removed from his consciousness everything that could be doubted, he discovered moral and social as well as mathematical rules, and identified those rules with the deity. Descartes discovered no moral or social rules. Such rules, like perceptual constancies, are subject to error (contain "nothingness") and whatever validity they have must be derivable from mathematics, which alone provides certainty.

Descartes's first step is to identify truth with whatever is clear and distinct in his own mind: "Since our ideas or notions have positive reality and proceed from God, insofar as they are clear and distinct, they must to this extent be true" (*Discourse,* part 4, A & G, p. 36). When all that could be doubted was discarded, what clearly and distinctly remained? "Nothing is more easily or manifestly perceptible to me than my own mind" (Second Meditation, A & G, p. 75). "*I* am a conscious being; that is, a being that doubts, asserts, denies, understands a few things, is ignorant of many, is willing or unwilling; and that has also imagination and sense" (Third Meditation, A & G, p. 76). Thus, the first class of clear and distinct mental contents is doubt and will; the second is imagination and sensation; the third is mathematical truth:

> I remember that even previously [to his effort to reject anything that could be doubted] at a time when I was utterly immersed in the objects of sensation, I regarded this kind of truth as the most certain of all—namely, those that I recognized as evident in regard to figures, and numbers, and other matters of arithmetic, or of geometry, or in general of pure abstract mathematics. (Fifth Meditation, A & G, pp. 102–103)

Descartes then attempts to prove the existence of God by virtue of the similarity of the concept of God to the concept of a mathematical truth:

> Now if it follows, from my mere ability to elicit the idea of some object from my consciousness, that all the properties that I clearly and distinctly perceive the object to have do really belong to it; could not this give rise to an argument by which the existence of God might be proved? I assuredly find in myself the idea of God—of a supremely perfect being—no less than the idea of a figure or a number; and I clearly and distinctly understand that everlasting existence belongs to his nature, no less than I can see that what I prove of some figure, or number, belongs to the nature of that figure or number. So, even if my meditations on previous days were not entirely true, yet I ought to hold the existence of God with at least the same degree of certainty as I have so far held mathematical truths. (Fifth Meditation, A & G, p. 103)

Unlike Augustine's internal God, Descartes's internal God is silent on morality—on how he ought to behave—except as regards mathematics.

This difference between the two philosophers, for the behaviorist, is the problem with the introspective method. Both Augustine and Descartes claim to have eliminated from their consciousness all except what cannot be doubted or what strictly follows from what cannot be doubted. Both philosophers claim that what remains requires only the will to see. Both claim that this undoubtable truth really exists as a perceptible object in the mind of every person; yet, when the method is applied, each discovers something different. Even the mathematical rules that they both claim to find in their consciousnesses are not the same. For Augustine, unshakable truths were simple arithmetic rules; for Descartes, unshakable mathematical truths were the common logical elements that he himself had previously discovered to underlie algebra and geometry.

It is not surprising that Descartes, who discovered a still more abstract set of rules embracing two different but already highly abstract systems should want to claim the status of true existence, true reality, for his discovery. The next step—the claim that (aside from his own conscious states) nothing else in the world has any existence or reality unless it is as clear to him as the mathematical reality he himself discovered—is harder to go along with.

Reality and Illusion

In his *Discourse on Method,* Descartes says: "I could take it as a general rule that whatever we conceive very clearly and very distinctly is true" (part 4, A & G, p. 32) and "I went through some of the simpler [geometrical] proofs, and observed that their high degree of certainty is founded merely on our conceiving them distinctly" (part 4, A & G, p. 34).

For Descartes, as for Plato, the concept of a triangle exists more certainly than does any particular triangle in the world. For both philosophers the concept is eternal, whereas particular triangles come and go. The difference between the two is that, for Plato, the concept consists of some abstract quality of the world and knowing the concept consists of an interaction (a functional one) between a person and that aspect of the world; for Descartes, the concept exists inside the person and knowing the concept is a private occurrence between the soul and the concept within the soul. The existence of triangles in the world outside of his own soul is, for Descartes, an inference from the fact (1) that we perceive triangles in nature and (2) that God would not deceive us: "It may be that not all bodies are such as my senses apprehend them, for this sensory apprehension is in many ways obscure and confused; but at any rate their nature must comprise whatever I clearly and distinctly understand—that is, whatever, generally considered, falls within the subject matter of pure mathematics" (Sixth Meditation, A & G, p. 116). Thus nothing can be said to exist with certainty in the world unless it be clearly and distinctly understood. What can be clearly understood about the world, in turn, is what can be reduced to pure mathematics.

For Aristotle, the reality of abstract entities was seen through particular objects in the world. Descartes reversed these priorities; for him, the reality of particular objects in the world was inferred through their abstract properties.

Although sensations were themselves clear to Descartes, he completely distrusts any inferences based upon them. Sensation is, for him, just a form of imagination that we are compelled to experience. We are more aware of reality when awake than asleep not because our *senses* are functioning when we are awake but because our *reason* is then functioning: "For if it happened even in sleep that one had some specifically distinct idea; if, for instance a geometer devised some new proof; then sleep would be no bar to its being true. . . . Waking or sleeping, we should never let ourselves be convinced except by the evidence of our reason" (*Discourse,* part 4, A & G, pp. 36–37).

Descartes and Behaviorism

If Descartes had looked inside himself and found *only* mathematical truth, he would have been forced to explain the existence of his own will (his "motivation") in terms of that mathematical truth; he might then have developed a mathematical theory of the mind not unlike mathematical theories of behavior currently being developed by cognitive scientists and behaviorists alike. Unfortunately, although the efficient causes of his own sensations (themselves already certain) were mistrusted by Descartes (therefore subject to mathematical confirmation), he could not mistrust the efficient cause of his own will, which he supposed to be God.

Had Descartes been able to consider his own will not in terms of its *efficient* cause but in terms of its (Aristotelian) *final* causes—the same (mathematically definable) objects in the environment that gave rise to his sensations—he might conceivably have been a behaviorist. But Descartes could not take this step because he had committed himself to a strong position (along with Renaissance physics) against final causes, against teleology. Thus, human motivation, for Descartes, was a wholly subjective event observable only by looking inward, prior to mathematics and outside the realm of science.

The One and The Many

The fundamental philosophical problem faced by Descartes in common with the other philosophers here discussed—the relationship between the abstract and the particular in human knowledge—may be viewed from a behavioristic perspective. Let us turn again to this issue, this time in its aesthetic form.

Consider a work of art that may be appreciated at more than one level, say, Beethoven's Ninth Symphony. Let us agree that this work has some individual elements (movements, sections of movements, themes, combinations of notes) that may be appreciated (understood, enjoyed) individually, and that the structure of the work as a whole may also be appreciated. The nature of a symphony, stretched out in time, makes its (particular) elements appreciable in a shorter time than its (more abstract) structure as a whole. We can talk of "concentration" or "focusing attention" on the particular elements. We can say that it is "tempting" to focus our attention on the elements because it is easy to do or because the pleasure of it is immediate.

What can we say if, on the other hand, we want someone to pay attention to the temporally extended structure of the work as a whole? We might say "concentrate on the structure," but the word "concentrate" implies a compression of attention (to a particular point) rather than an expansion (to an abstraction). However, if we conceive of the structure as an unextended point within our own bodies, then the metaphor becomes appropriate. The particular theme of the symphony thus seems to exist outside of us and to be directly appreciated and concentrated on in the ordinary sense, while the abstract structure of the symphony seems to exist inside of us (in our memory) to be appreciated in a different area of concentration. Attention to particular or abstract aspects of the symphony would then be conceivable as shifted (voluntarily) between the external particular point and the internal abstract point, as a searchlight is shifted from one point to another. This Augustinian and Cartesian conception of the relationship between knowledge of the particular and the abstract differs from the Platonic conception where the difference between attention to the particular and the abstract was a difference in the *kind* of attention rather than a difference in the place of attention.

Had Descartes firmly kept to this strict dualism: abstract inside, particular outside, it would have not been difficult to translate Cartesian philosophy into behavioristic terms. The difference between (1) "Joe Jones's act is controlled by relatively abstract aspects of the environment" and (2) "Joe Jones's act is caused by his mind [or soul]" is only a metaphorical or semantic difference if the mind (or soul) is conceived as the repository of the more abstract aspects of the environment. (The further question, Is the capacity of a person's behavior to be so controlled or so caused innate or learned? is orthogonal to the question of whether (1) or (2) is the more useful expression.)

However, despite the modern understanding of Cartesianism as a dualism of the abstract and particular, Descartes believed that his own sensations (as opposed to their efficient causes), his own imaginations, his own motivations were as "clear and distinct" as mathematical rules, not themselves subject to mathematical analysis and therefore not subject to scientific understanding. It is this confusion of terms (mathematics, will, and sensation, all on the same epistemological level) that makes Descartes impossible to reconcile with behaviorism.

After Descartes

From Descartes to the present day the main question in the philosophy of mind has been: What is clear and distinct and what may be derived from it? Descartes believed that the existence of objects in the world could be proved through the application of one clear and distinct entity (mathematical rules or reason) upon another (sensations). It seemed obvious to subsequent philosophers that the epistemological status of reason was somehow different from that of sensations, although both might be clear and distinct. Kant's elegant solution was to separate phenomenal experience in all its complexity from the mind's a priori categories (including reason), on one hand, and from the structure of the physical world, on the other. Phenomenal experience was then viewed as the (efficiently caused) outcome of the interaction between the world and the mind, while the world and the mind could be inferred to exist on the basis of phenomenal experience. The job of psychology (taken up by the gestalt psychologists; see Chapter 1) would then be to analyze phenomenal experience so as to better and better construct a world and a mind whose interaction would have given rise to just that experience.

For John Stuart Mill, only elementary sensations themselves were clear and distinct (in Descartes's sense) in the mind. The existence of the physical world could be derived from those sensations together with (chemical-like) laws of their association. Logic (reason, mathematics), for Mill, is nothing but the action of these laws. The job of psychology (taken up by Wundt and Titchener, as described in Chapter 1) would then be to discover the elements and the laws of their combination.

For both Kant and Mill, psychological observation is internal and private to the individual. For Kant, each person's phenomenal experience is his or her own, although Kant, like Augustine (see Figure 6.1), conceives of the pair of eventual constructions from it (both sides of Figure 6.1) as common among human beings. For Mill, the common aspects of the mental world are completely abandoned; what each person sees in his or her mind is private to that person. Only the laws of association and the world as the source of sensations are common among people.

It was not until the present century that the "clear and distinct" aspects of a person's mind were again conceived (as they had been by both Plato and Aristotle) as observable in that person's overt behavior.

Notes

1. Unless otherwise indicated, the quotations from Augustine are from *Augustine: Earlier Writings,* edited and translated by J. H. S. Burleigh (Philadelphia: Westminster, 1953).

2. Or at least from the Neoplatonic amalgam of Eastern mysticism and Platonic philosophy.

3. W. V. O. Quine's (1960) concept of "opacity" of mental terms reflects Augustine's usage. The statement "I believe that Mark Twain is a good writer" is not the same as "I believe that Samuel Clemens is a good writer," if I do not know that Mark Twain is Samuel Clemens *and* if the statement is thought of as privileged evidence of my belief. But if the statement were merely one bit of behavior constituting my belief (as behaviorists say), then the opacity becomes trivial. The truth of the statement, for a behaviorist, depends on its correlation with other behaviors. Thus, (1) Person X says, "I believe that Mark Twain is a good writer," and (2) Person X reads a lot of Mark Twain, may be conceived as mutually corroborating bits of evidence regarding Person X's belief. For a behaviorist the opacity of "Mark Twain" in the first bit of evidence is (trivially) due to its enclosure within quotation

marks and not at all an indication of an internal state—even when Person X is oneself.

4. The notion that all earthly desires are bad because they can be lost and the identification of happiness as the attainment of something good that cannot be taken away is a characteristic of Augustine's philosophy that it (and Christian belief generally) shares with Stoicism, albeit the two philosophies are based on vastly different metaphysics.

5. All Descartes quotations are from *Descartes' Philosophical Writings,* edited and translated by E. Anscombe and P. T. Geach (Indianapolis, IN: Bobbs-Merrill, 1971). Because the various works have no paragraph numbering, citations are to page number in this text (abbreviated A & G).

6. The Catholic church was not much happier with this division than they would have been had Descartes categorized all human behavior as involuntary (subject to empirical study). Natural philosophers could be said to know better than priests one aspect of God (creator of the material world), but at least there were only a few natural philosophers. The second category of human behavior, voluntary behavior, implied that each individual person (of which there are many) could by internal reflection find another aspect of God inside him. What remained for Catholic authority was only that authority itself, unsupported by superior knowledge of God. This must have seemed too close to Protestantism. Descartes, however, seemed to think his philosophy less dangerous than his psychology because during his lifetime his philosophical writings were published—not without misgivings.

7

A Teleological Science of Psychology

The crucial event for psychology between René Descartes and modern times was the publication of Charles Darwin's (1809–1882) theory of natural selection, in his *Origin of Species* (1859). If the human body evolved to better and better fit into the environment, the human mind (reasoned Darwin and his followers) must have evolved in a corresponding way; both body and mind might be understood (as Aristotle originally suggested) in terms of their functions in the environment. Twentieth-century philosophers, especially Gilbert Ryle and Ludwig Wittgenstein (1889–1951), have noted that the relationship between people's minds and the environment cannot be conceived separately from their overt behavior. The focus of the study of mind has thus shifted in our century from introspection to behavior. This change of focus, however, has effectively split psychology off from philosophy. The crucial issues in understanding the human mind have become the psychological issues delineated in Chapters 1 and 2 rather than one or another side of a philosophical argument.

Nevertheless, the distinction, illustrated in Figure 2.2, between a (cognitive or physiological) science of mental mechanisms and a (teleological behaviorist) science of mental life has brought psychology to a critical point. The question arises whether it is at all meaningful to study the human mind in terms of final causes. That question will be answered ultimately by the success of one kind of psychology or another in the treatment of mental illness, the development of effective educational methods, the control of addiction and crime, and other human problems. Since, obviously, there has as yet been no decisive success in these areas, psychologists, before plunging ahead, might well stop and think about the assumptions underlying their research. This final chapter, therefore, skips over the philosophical debates about the nature of mind between Descartes and the present decade and goes straight to the current issue of the path psychology ought to take.[1]

Efficient and Final Causes in Modern Psychology

The success of Renaissance physics based on efficient causes has led psychologists to attempt to discover the efficient causes of behavior in cases where no current efficient causes can be found in the environment. Such behavior has been classified by philosophers from Plato to Descartes as voluntary. If not in the environment, most modern psychologists have reasoned, following Descartes, those causes must be within the person, either in the mind (according to cognitive psychologists) or in the nervous system (according to physiological psychologists).

Most modern cognitive theories in psychology attempt to discover the "internal representations" of environmental events and to deduce the interactions and processing of those representations so as to explain a person's behavior. While verbal report based on introspection is no longer considered to be a valid prima facie reflection of internal states (Sigmund Freud (1856–1939) among others convincingly demonstrated the unreliability of what people say about the states of their own minds), what a person says, like what a person does, serves as data for all modern psychological theories, behavioral as well as cognitive. Cognitive theories construct their models of the mind from verbal reports plus overt behavior. Although introspection has lost its privileged status, the object of most modern cognitive theories, no less than that of Descartes's theory, is to discover the internal efficient causes of human behavior.

Skinner's (1938) concept of the operant—behavior defined in terms of its consequences—and his method of operant conditioning provide the groundwork for another kind of science of voluntary behavior, one based on the Aristotelian concept of final causes. While psychologists, including most behaviorists, have emulated physics in rejecting final-cause explanations of behavior,[2] twentieth-century physicists (in a sort of end run) seem to have accepted final causes as well as efficient causes as valid explanations of the behavior of physical objects (see the quotation from Max Planck, p. 23). Because the science of final causes was developed by Aristotle on the basis of his conception of the structure of human thought, a behavioristic science of final causes might be expected to be at least as well suited for the study of psychology as one based on efficient causes.

Modern philosophers are skeptical whether such a psychology would be meaningful or is even conceivable. John R. Searle (1981, p. 405) says: "Most philosophers today regard behaviorism as dead." But, as I shall now argue, their arguments are valid only against a molecular conception of behaviorism based on efficient causes. Other philosophers, such as Daniel

Dennett (see Chapter 3, note 6), regard behaviorism as unscientific or not really psychology. The behavioristic interpretation of Aristotle's psychology in Chapter 5 is an attempt to demonstrate that the understanding of the human mind achievable by a teleological study of the behavior of whole organisms may be as rich and as useful as that achievable by the methods of Renaissance science.

Current Philosophical Objections to Behaviorism

The two sorts of mental states that were clear and distinct to Descartes correspond to two dimensions on which, according to the contemporary philosopher Richard Rorty (1979, p. 24), philosophers distinguish between mental and physical states: intentional–nonintentional and phenomenal–nonphenomenal. Examples of intentional states are "beliefs, desires, intentions." Examples of phenomenal states are "raw feels—e.g., pains and what babies have when they see colored objects."

Gilbert Ryle (1949) argued that intentional mental terms such as "belief," "hope," and "knowledge" refer not to discrete events but to behavioral *dispositions*. A disposition (like the disposition, meant by the word "brittle," of a cup to break when dropped) is not a particular event. Philosophers (like Descartes) who conceive of mental terms as particular events inside the body are, according to Ryle, making a category mistake. A category mistake is the use of a term for a class of events as if it could stand alongside a member of that class (like the boy who, when given a poodle, wants a dog also). For Ryle, intentional mental terms stand for classes of particular overt actions. Ryle is therefore a teleological behaviorist.[3]

P. T. Geach (1957), translator of Descartes, says in criticism of Ryle that explanation of "a statement of an *actual* difference between two men's mental states as really asserting only that there are circumstances in which one *would* act differently from the other . . . could be *all* the difference there is between the two [runs counter] to a very deep rooted way of thinking. When two agents differ in their behavior, we look for some actual, not merely hypothetical, difference between them to account for this" (p. 5). For instance, if I say, "I am worried," and you say, "I am not worried," the difference between me and you is not in what we would do if things were different (a merely dispositional distinction) but in something we are doing right now: I am worrying and you are not. But it is not clear that Ryle meant dispositions to be so ontologically weak. In terms of Chapter 5's discussion of first and second degrees of actuality (see Figure 5.1) Ryle may be seen as claiming that all mental states (such as worrying) may be actual in the first degree as well as the second. It makes perfect sense for me to say, "Nahid [my wife] is worried that I don't have enough life insurance," in her absence or even while she is asleep. Her worry would be no less real or actual for that. Her worry (as actual in the first degree) consists of the overall pattern of her frowns, tones of voice, verbal expressions, and physiological states, a pattern that might extend over weeks or months. This pattern forms the context for each of the particular events (frowns, and so forth) of which it consists (as a jigsaw puzzle forms the context of each of its pieces or as a symphony forms the context of each of its notes). Just as we normally conceive an orchestra to be playing a symphony even between movements while it is actually (in the second degree) not playing a note, so my wife may be worrying even while asleep. Her particular frowns (and so forth) are, according to Aristotle, actual only in the second degree. There is no reason to suppose that an abstract, categorical, or temporally extended environmental event is less "real" than a concrete, particular, temporally narrow environmental event (for Plato, abstract events were the *only* reality). But regardless of what Ryle meant, Geach's argument has no force against teleological behaviorism; that is, a view of a person's mental states in terms of *actual* overt behavior caused by *actual* environmental contingencies.[4] (Furthermore, it is not at all clear what Geach means by "a very deep rooted way of thinking." By Geach's own argument, a way of thinking cannot mean a common way of speaking or acting; it would seem that depth of thought means to Geach merely what is clear and distinct to him upon his own introspection.)

It is not possible here to discuss in detail the arguments of every one of the "most philosophers" who regard behaviorism as dead. Instead, I shall focus on three: Ned Block (1981) and John R. Searle (1980), who claim that behaviorism cannot account for intentional states; and Hilary Putnam (1980), who claims that behaviorism cannot account for phenomenal states. I shall also show that both sorts of objections to behaviorism have force only when behavior is conceived in terms of discrete movements over a brief time span, not when behavior is conceived as a long-term pattern of interaction of a whole organism with its environment.[5]

Intentional States

Block (1981) argues against behaviorism and in favor of what he calls *psychologism*, which he defines as follows (p. 5):

> Let psychologism be the doctrine that whether behavior is intelligent behavior depends on the character of the internal information processing that produces it. More specifically, I mean psychologism to involve the doctrine that two systems could have actual and potential behavior *typical* of familiar intelligent beings, that the two systems could be exactly alike in their actual and potential behavior, and in their behavioral dispositions and capacities and counterfactual behavioral properties (i.e., what behaviors, behavioral dispositions, and behavioral capacities they would have exhibited had their stimuli differed)— the two systems could be alike in all these ways, yet there could be a difference in the information processing that mediates their stimuli and responses that determines that one is not at all intelligent while the other is fully intelligent.

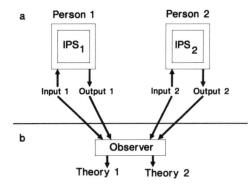

Figure 7.1 Two people with two different information-processing systems (IPS_1 and IPS_2) each (a) with its own input and output. (b) An observer constructs two theories (Theory 1 and Theory 2) to explain relationships between inputs and outputs.

Figure 7.1a (my own figure, not Block's) illustrates the point. Computers and human beings share the property of indeterminacy: that a single finite pattern of behavior, no matter how complicated or extensive, may be equally well explained by more than one theory. For Block, a theory of intelligence is an "information-processing system." Even if information-processing systems (IPSs) are only a subclass of possible theories of behavior, it is true that any given behavioral pattern can be explained by more than one information-processing system. Referring to Figure 7.1a, given Input 1 = Input 2 and Output 1 = Output 2, it still may be true that $IPS_1 \neq IPS_2$. If a single set of inputs and outputs can be accounted for by both an IPS_1 that observers would agree is not intelligent and an IPS_2 that observers would agree is intelligent, then it must be the IPSs to which observers refer when they talk about intelligence, not the inputs and outputs.

Block's argument is a more formal version of Searle's (1980) "Chinese room" parable. Searle imagines a non-Chinese speaker in a room with a list of Chinese characters and an instruction book telling how to arrange them in all conceivable sequences. A questioner passes a list (a question) through the door. The man by looking into his instruction book, completes the sequence (answers the question) so well that it is impossible for the questioner to detect any difference between the man's responses and that of a person who actually speaks Chinese. Yet the man does not speak Chinese.

The negative part of Block's and Searle's argument is correct. Mere symbolic manipulation can never capture the essence of what we mean by a mental capacity such as intelligence. As Searle (1989) says, such a system, like his Chinese room, deals only in grammar, not in meaning: "just having a formal syntax is not sufficient for understanding; syntax is not semantics."

Both Block and Searle imply that by satisfying Block's input-output test for intelligence a computer would be behaving exactly like an intelligent human being. But there are many differences between the behavior of such a computer and that of a human being, and it is precisely by virtue of those differences that we say the machine is not intelligent.

For a behaviorist the meaning that computation normally lacks (the thing that would make computation intelligent) is intelligent overt behavior—intelligent interaction with the environment. Searle (1989) considers this possibility and rejects it. Why? As an imaginative exercise Searle extends his Chinese room argument. He supposes himself (a non-speaker of Chinese) to possess "a rule book for manipulating Chinese symbols according to computer rules." By using the rule book his "behavior is indistinguishable from that of a native Chinese speaker" and, furthermore, everyone is convinced that Searle understands Chinese. Let us then imagine Searle walking down a street of Peking. Stopping a passerby, he asks directions to his hotel in fluent Chinese. He is given directions in Chinese (not a word of English is spoken nor is a gesture made). He then heads straight for the hotel. As he is about to cross the street another passerby yells something in Chinese and Searle jumps back onto the curb just as a bus rolls past. He has nearly lost his life, but thanks to that book in his pocket and his speedy reading of it his life is saved. (No one sees him consult it; otherwise, they might not be convinced that he knows Chinese. Perhaps the book is written in Braille and he quickly reads it with his fingers.) Then Searle enters the hotel, registers without a hitch, meets a delegation of Chinese philosophers, and talks to them in Chinese. Later he writes a paper in English and they, in Chinese, which remarkably agree about the nature of the complex and highly detailed conversation. This is the sort of behavior, "indistinguishable from that of a native Chinese speaker," that would convince everyone that Searle understands Chinese. Of course, everyone would be convinced—and rightly so. The meaning of "knowing Chinese" is using Chinese effectively in the world. A computer cannot understand Chinese because a computer cannot by itself use the Chinese language as a native Chinese speaker does.

Block's argument follows the same lines as Searle's. He describes an unintelligent IPS that produces behavior that could also have been caused by an intelligent IPS. The unintelligent IPS is created by a group of superprogrammers who anticipate all the conceivable inputs to the machine and the conceivable outputs that an intelligent person might make given those inputs. Block argues that this set of inputs and outputs (called a set of "sensible strings") is finite. Let us grant this point.

The programmers then simply record all possible sensible strings and load them into the computer's vast but finite memory. The computer can then interact with people just as though it were intelligent. Yet, Block says (p. 21), "the machine has the intelligence of a toaster." People interacting with the machine would suppose the machine to be intelligent but they would be fooled. If they really knew how the machine was programmed, they would admit their error. Block says (p. 27), "though we *normally* ascertain the intelligence of a system by trying to assess its input-output capacities, it is part of our ordinary concept of intelligence that input-output capacities can be misleading."

People probably would indeed agree that the computer Block describes could be unintelligent—but not for the reason Block supposes. The computer in Block's example deals with inputs and outputs in the form of sentences. It is only after describing that kind of input and output, and asking for the reader's agreement that *that* computer may be unintelligent, that Block extends the notion of input and output to include the normal environment (in addition to sentences) with which intelligent human beings ordinarily deal and the normal behavior (in addition to sentences) they emit. Then, after introducing this extension, in fending off potential objections to his argument, Block retreats again to the purely linguistic computer example. But the aspect of the computer's behavior that acts on the reader's intuition and compels agreement that the computer *may* be unintelligent despite its behavior is not anything about the way it is programmed. The computer appears unintelligent because of (1) the fact that its inputs and outputs are imagined only as slips of paper with writing on them and (2) Block's failure to specify any function for the machine.

Consider, therefore, another computer, one that looks something like a person, something (perhaps, like Dolly II of Chapter 1) programmed as Block supposes by superprogrammers. Suppose the superprogrammers programmed these computers to help humans in return for occasional oiling so that a "sensible string" might be: (1) "Please oil me—I'll do anything you want." (2) "No, I don't think I will." (3) "Please, I beg you. If you do, I'll help you with your homework." (4) "O.K." [At this point oil is poured into a hole in the computer.] (5) "Thank you dear, now let's see that homework, and as a special treat, I'll sing you a song afterward, and give you a nice hug and kiss before you go to sleep," and so forth. Then the computer proceeds to do what it has promised to do. Block, in describing the programmers' job, says (p. 20):

> Presumably the programmers will find that in order to produce really convincing sensible strings, they will have to think of themselves as simulating some definite personality with some definite history. They might choose to give the responses my Aunt Bertha might give if she were brought to a room with a teletype by her errant nephew and asked to answer "silly" questions for a time.

But what may be required to give most people a (clear and distinct) idea of an intelligent machine is that Aunt Bertha's behavior *outside* of the room be simulated, including her kisses, hugs, birthday presents, expectations of birthday cards from Block, feuds with Block's mother, production of chicken soup, and so on. That behavior is what Aunt Bertha's intelligence consists of, in addition to her answers for a limited time to a few "silly" questions. Accepting the proposition that a group of programmers could program, in a finite time, Aunt Bertha's nonverbal as well as verbal behavior according to Block's specifications, who would not (clearly and distinctly) see an Aunt Bertha machine as intelligent?

Suppose some religious cult says that God created us just as the programmers are said to have done. (These cult members say, if complimented on their intelligence, "Oh no, it was not me, it was God.") Would we treat them like toasters?

Suppose two hostile armies were approaching a city, one composed of what are known to be robots programmed by Block's superprogrammers and one composed of normal human beings. The general in command of the city's defense wants to announce to his troops that both of these armies must be feared because of their clever tactics, their superior use of weapons, their excellence in camouflage, their trickiness. Would it not be a complete corruption of "our ordinary use of intelligence" for the general to say that one army is intelligent and one is not?

If, as John Dewey (1896) claimed, the terms "stimulus" and "response" are not "distinctions of existence" but "distinctions of function"; and if, as Wittgenstein (1958) imagined, the very simplest conceivable language involved functional interchange, how could our "ordinary use of intelligence" be a label for a computer program, hidden wholly within the organism, with no apparent function?

Another way of considering intelligence is shown in Figure 7.1b. In this model an observer considers the inputs and outputs of two systems, notices relationships between the inputs and outputs of each system, and constructs a theory, the sole purpose of which is to enable the theorist to predict future outputs on the basis of future inputs. It may be that Input 1 = Input 2 and Output 1 = Output 2, yet Theory 1 $\neq$ Theory 2. Let us say that the observer is a cognitive psychologist and that the two theories are two alternative hypotheses about the computer's actual program. It may be that Theory 1 contains a variable that the observer calls intelligence, whereas Theory 2 contains no such variable. It would be foolish, however, for the observer to claim that she has somehow tapped the "ordinary use of intelligence" in such a way that a person stimulated by Input 1 and producing Output 1 is intelligent, whereas a person stimulated by Input 2 and producing Output 2 is not. It is obvious that a person behind Input 1 and Output 1 could just as well have been the person behind Input 2 and Output 2 since, by hypothesis, these inputs and outputs are identical.

Let us suppose that the observer is the viewer of a play and that Theory 1 is the theory that the actor playing the part of a brilliant lawyer is really intelligent, while Theory 2 is the theory that the same actor is a moron, just mouthing intelligent lines. Both theories are consistent with the data and both theories are about the actor himself. If the playgoer adopts Theory 1 but switches to Theory 2 when the actor is viewed in the context of everyday life, it would *not* be true that the actor's intelligence depends on how gullible the playgoer was. The actor's intelligence depends, rather, on his behavior other than that immediately available to the playgoer. Whether the actor is *really* intelligent depends on what sort of (offstage) behavior is considered to be intelligent and on whether the actor's behavior conforms to it.

Current computers do not think for the same reason that they do not feel pain or possess other mental states: their overt behavior—including their interaction with human beings—is all too distinguishable from that of real human beings. The day when a computer is built that does indeed pass this test will be the day that we all agree (Block's and Searle's protestations notwithstanding) that computers have minds.

A behavioral analysis of intelligence might begin with its primary reinforcement: how intelligence preserves people in their environments. It might proceed to more abstract (secondary) reinforcers, such as the social consequences of intelligence, then to the study of intelligence for its own sake, and finally to its structure (its syntax). There is plenty to do here, in addition to the equally valid and essential (but not essentially psychological) investigation of cognition and physiology. The process of designing a truly intelligent computer would have to start with the establishment of behavioral criteria. Only then could one proceed to the design of internal structure.

Phenomenal States

If you take a cognitive or physiological viewpoint from which to consider intentional mental states, then there will seem to be an unbridgeable difference between an intentional mental state, such as belief or hope (describable from a cognitive or physiological viewpoint in terms of interactions among internal mechanisms), and "raw feels," such as pains and what babies feel when they see colors.[6] No matter how detailed or complete or physiologically correct are the plans of the cognitive or physiological mechanism hypothesized to underlie pain (for instance), there is something about pain that is in principle uncapturable by those plans.

This point was made forcefully by Dennett (1978, pp. 190–229), who speculated on how elaborate a computer would have to be to be able to feel pain. After taking more and more factors into account (the nature of the painful stimulus, its emotional context, its cognitive context, and various interactions thereof), Dennett finally admitted that none of the elaborations of the program would change the mind of human critics, who stubbornly maintained that the computer just could not feel pain as they did. Dennett's point is that "raw feels" such as pain, as they are commonly conceived, have no psychological meaning. He implies that pain as well as other "raw feels" are most usefully studied as intentional systems (cognitive or physiological systems) and that psychologists ought simply to forget about trying to account for "raw feels" in a way that will capture their meaning in everyday speech. But the reader who has gotten this far will recognize that Dennett is identifying scientific respectability with efficient causation. "Raw feels" have identifiable functional roles (identifiable final causes) in everyday life.

A wagon with a squeaky wheel is more likely to be seen as a machine in pain than a computer programmed as Dennett supposes. From a behavioral viewpoint, the wagon with the squeaky wheel seems more in pain than the computer because of certain functional similarities between the wagon and a person in pain—namely, the relationship between the intensity and frequency of the squeak and damage to the wagon, and the ability of the listener (by oiling the wheel) to reduce the squeak and the damage simultaneously. (Of course, there remains a vast difference between squeaking wagons and people feeling pain.)

A behaviorist finds phenomenal mental states, like intentional mental states, in the interaction of a whole organism (including oneself) with the environment—in behavior. But because there seems (from other viewpoints) to be such a vast difference between intentional and phenomenal states, it is worth attempting to sketch a teleological account of pain, a phenomenal state, the self-evident nonbehavioral existence of which has often been said to prove behaviorism wrong.

An influential critique of behaviorism on these grounds is the argument of Hilary Putnam (1980). Putnam asks us to imagine a community of super-Spartans who are trained from birth to inhibit all pain behavior except that which is verbal and, even when they say they are in pain, to say it in a calm voice. Using our previous terminology, super-Spartans, by virtue of childhood training, have learned to mold their responses to painful stimuli into a form (the words "I have a pain in my toe" spoken in a calm voice) that is useful for functions such as telling a doctor where it hurts but not useful for functions such as stopping a baseball game (clutching and falling), warning other people of danger (yelling loudly), or avoiding tissue damage (pulling one's hand quickly from the fire). In the super-Spartans' world, "psychological" pain would not be a problem (as it is in our world), but they would pay a price in burned hands for their lack of it.

Super-Spartans calmly *say* they are in pain but do not exhibit any other pain behavior. We can certainly conceive of such people and imagine how they might have been trained. But does that first conception necessarily entail a second and quite separate conception: that super-Spartans *have* pain exactly when they say they do? Putnam's argument against behaviorism hinges on the second conception. But there is a good deal of evidence (reviewed by Nisbett & Wilson, 1977) that what people say about their internal states does not correspond to what the best available cognitive theories (theories that explain behavior in a wider context) claim that their internal states actually are. (And behavioral theories insist even more forcefully on this distinction. W. E. Fordyce (1988), the director of one of the most successful clinics for chronic pain in the United States, has found it necessary, in pain assessment, to distinguish sharply between what people say about their pain and what they do.) How *should* one interpret what super-Spartans say? Consider the following example.

Let us suppose that three super-Spartan warriors visit a super-Spartan doctor the day after a battle. All three claim (in calm voices) to be in excruciating pain. The pain descriptions of two of the warriors correspond precisely to their wounds. The first says he has a "throbbing pain" in his right shoulder, and sure enough, one kind of wound (say, a large bruise) is found on his right shoulder; the second says he has a "burning pain" on his left thigh, and sure enough, another kind of wound (say, a deep cut) is found on his left thigh. If all pain reports were of this kind—overt verbal behavior correlated with identifiable tissue damage—then the warriors' pain reports could be seen as discriminative stimuli emitted to guide the doctor's behavior; the reports would serve the same function as saying, "Pass the salt," when the saltshaker is out of your reach but in someone else's. The doctor would then treat the wound. The words in which the soldiers describe their pain could then be understood like the words "pass the salt"—as guides to behavior. The doctor (and the super-Spartan philosopher) need only deal with the words as helpful descriptions of the wounds. The conception of a mediating internal state called pain would (for these two warriors) be superfluous. The words "throbbing" and "burning" could just as well be replaced by "Type-A" and "Type-B."

The third warrior, however, has what is usually called a psychological pain. He says (in an equally calm voice): "I have a terrible pain floating around inside me"; however, neither a wound nor any other sign of physical injury is found. The doctor might then suggest changes in the childhood training of super-Spartans so as to avoid such wastes of medical attention in the future, or just shrug his shoulders and dismiss his patient. As a super-Spartan himself he would not show any sympathy to the warrior or give him a day off. What to us seems cruel would, to a super-Spartan, be the normal behavior, cruel or not, of his compatriots. Otherwise, we would have to conceive a kind of childhood training that compelled super-Spartans to inhibit pain behavior even though that behavior was reinforced in super-Spartan society. If super-Sparta entails suspension of the normal laws of human learning as well as normal pain behavior, it becomes harder to envision. Within the limitations of childhood training, as we know it, there is nothing about super-Spartan behavior that compels abandonment of the teleological viewpoint.

Sensing this, perhaps, Putnam imagines the evolution of super-Spartans into *super*-super-Spartans. The super-super-Spartans never even *say* they are in pain, and they need no childhood training. They act this way from birth. There are numerous problems in conceiving of super-super-Spartans. First, there is no way (given Darwinian evolution) for super-super-Spartans to evolve. While it is conceivable (however unlikely) that pain behavior should be so maladaptive that people exhibiting the slightest trace of it (let us call them "crybabies") would die an early death (prior to reproduction), there would be no way to distinguish the super-

super-Spartans who inhibit all pain behavior from others who are completely insensitive to pain (let us call them "numb ones"). By hypothesis, neither nature nor super-Spartan kindergarten teachers could selectively kill off numb ones and still allow super-super-Spartan children to survive. But let us put aside this puzzle and try our best to conceive of super-super-Spartans. (If they could not evolve, perhaps they could be built.)

Suppose that a super-super-Spartan genius discovers the neurological substrate of the pain that the super-super-Spartans always inhibit. (Putnam says that super-super-Spartans still have internal physiological pain.) Perhaps the connections from the pain mechanism to the motor apparatus have atrophied in super-super-Spartans. Suppose that the genius realizes what a great advantage in terms of tissue-damage avoidance it would be if super-super-Spartans could communicate to one another when the pain mechanism (which from his point of view could only be a more or less accurate internal indicator of tissue damage) was active. The genius then devises a red light that can be mounted on people's heads, the intensity of which reveals the output of this internal mechanism. Let us say that all super-super-Spartans were fitted with this red light from birth. Now, a super-super-Spartan baseball game could stop if a player's red light were lit; super-super-Spartans would be sympathetic and pay lots of attention to other super-super-Spartans whose red lights were frequently shining; super-super-Spartans would have days off from work when their red light was on; in severe cases, they would receive narcotics and other drugs that dimmed the red light and incidentally made them feel good; and so on. How long would it be, do you suppose, until super-super-Spartans would have to establish clinics to deal with excessive red-light shining? Put the overt *function* back in the pain and the super-super-Spartan world would be our world, with this difference: that the behavior we call pain, they would call shining one's red light.

Perhaps, eventually, the super-super-Spartans would learn to inhibit the shining of the red light on certain occasions even though their internal mechanism was active. They might then come to speak of a red-light shining within them even though none was shining outside. But this would be just a metaphor. There might emerge, in super-super-Sparta, two ways of studying red-light shining: (1) the external factors—the stimuli, rewards, and punishments that control it; and (2) the internal physiological-cognitive mechanism discovered by the super-super-Spartan genius plus the other mechanism (acquired by super-super-Spartans since the death of the genius) by which the output of the first mechanism was inhibited or augmented so as to control the red light. (These correspond to our own external study of pain behavior and internal study of pain mechanisms.) A *third* discipline, initiated by super-super-Spartan mentalists, the study of internal red-light shining based on introspective reports, would be pointless. Introspective reports of red-light shining could serve as data for super-super-Spartan behavioral studies (which might attempt to discover the reinforcers of such reports) or for super-super-Spartan physiological-cognitive studies (which might attempt to discover the internal mechanism by which such reports were generated). The *content* of the introspective reports, however, their testimony as to the existence and the nature of the internal red light, *is no evidence whatever* either for the existence or for the nature of an internal red light, which, as we have already said, would be a metaphor—not only according to behavioral theory but according to physiological and cognitive theories of red-light shining as well.

A Behavioral Account of Pain

Putnam is not alone in his use of "pain" as an example of a term not amenable to behavioral analysis. Whether it is called an emotion, a feeling, or a sensation, pain is often thought to be so self-evidently nonbehavioral that its very existence constitutes evidence against behaviorism. Recall, also, Rorty's (1979) classification of mental terms as intentional or phenomenal. Pain is the most common example of a state said to be mental on phenomenal grounds. There is some tendency, among philosophers, to use pain with respect to behaviorism as one might use a cross with respect to a vampire. Searle (1980) says: "Are there no pains underlying Rachlin's pain behavior? For my own case I must confess that there unfortunately often are pains underlying my pain behavior, and I therefore conclude that Rachlin's form of behaviorism is not generally true." Even Wittgenstein (1958, no. 304) seems to accept the internal reality of pain: " 'But you will surely admit that there is a difference between pain-behavior accompanied by pain and pain-behavior without any pain?'—Admit it? What greater difference could there be?"[7]

It may be worthwhile, therefore, to discuss pain briefly apart from its use as an illustrative term in antibehaviorist philosophical arguments. The following is far from a full-fledged behaviorist account of pain, but merely an illustration of a behavioristic approach to a topic, apparently intractable to such an account.

What does it mean to say that pain is just pain behavior? It means the same sort of thing as to say that a pigeon's pecks are just pecking behavior. Thus (as there certainly may be a physiological theory of pecking) it does not mean that a physiological theory of pain (endorphins and all) is impossible or not a worthwhile undertaking. It does not even mean that a cognitive theory of pain, in which pain would be an internal state interacting with other internal states, is impossible. It does mean that a behavioral theory is possible in which "pain" stands for a pattern or a set of patterns of overt behavior organized in terms of their consequences (final causes) as well as in terms of their antecedents (efficient causes).

A necessary preliminary step to a behavioral theory of pain would be to show how pain, like pecking, may be an "operant" (a pattern of behavior manipulable by its consequences) as well as a "respondent" (a pattern of behavior manipulable by its antecedents). As an illustration of how an act, usually considered a purely behavioral act, is learned, consider the development of eating.

According to P. Teitelbaum (1977), the patterns of learned behavior, over the lifetimes of animals, themselves fall into a larger pattern. At birth most behavior is strongly correlated with an immediately prior environmental stimulus (in Skinner's terms, is "respondent"). As an animal develops, however, the strong correlations to immediately prior environmental stimuli (efficient causes) loosen; behavior comes to be less and less strongly correlated with any particular prior stimulus but able to be reorganized by patterns of reward (subsequent environmental stimuli). For instance, at birth, eating by mammals tends to be relatively rigidly determined (turning to a nipple and sucking). As a mammal develops, it comes to eat semiliquid and finally solid foods, which require more instrumental preparation. The later phases of development are more varied and more subject to alteration with environmental conditions. Eating, in its final form, may have few elements in common with the original, largely respondent eating, but aspects of that earlier behavior may reappear and even dominate at moments of stress (Breland and Breland, 1961). After traumatic brain surgery or stroke, the original respondent pattern may occur in toto.

Instances of eating by an infant, according to Teitelbaum, are highly correlated with (efficiently caused by) specific prior environmental events, relatively invariant, and brief. An adult's eating is more loosely correlated with (prior and subsequent) environmental events, relatively variable and longer in duration. The eating of an adult mammal may bear only superficial resemblance to the eating of a newborn mammal. Eating, as it develops, becomes operant behavior (controlled by its consequences) as well as respondent behavior (controlled by its antecedents).

Like eating, pain begins as a relatively fixed respondent—a pattern of behavior correlated with immediately preceding stimuli. It becomes, in adults, a much more variable instrumental act—a pattern of behavior controlled by its (often social) consequences as well as by preceding stimuli. As with eating, the later pattern only partially overlaps the earlier one. It may seem that pain (behavior or not) is a fixed thing unaffected by normal rewards and punishments. But an overwhelming amount of evidence indicates the contrary. To someone not previously familiar with the literature on pain, the wide lability of pain depending on "psychological" factors may come as something of a surprise. The painfulness of a stimulus can be increased or decreased over wide ranges by such factors as placebos (McGlashen, Evans, & Orne, 1969), hypnotism (Hilgard & Hilgard, 1975), acupuncture (Melzack, 1973), sociocultural factors (Tursky & Sternbach, 1967), and the setting under which the stimulus was inflicted (Beecher, 1956). These factors are often so strong that they completely override the presence or absence of normally painful stimuli: "It can be said with certainty that psychological factors quite often cause pain and frequently augment its severity. They may also serve to abate or abolish it even in the presence of extensive trauma" (Merskey, 1968, p. 297). Some

Notes

1. I have not discussed one movement in the philosophy of science that was influential in the first half of the twentieth century: logical positivism. Logical positivism seemed at first to have much to offer to psychologists. By its claim that all science could be reduced ultimately to "sense data" and their logical relations, logical positivism seemed, to some psychologists, especially those studying sensation, to make psychology the propaedeutic science (Stevens, 1936).

Many behaviorists were also attracted to logical positivism (for example, Brunswik, 1952) because it seemed to offer an objective way of speaking about the mind and to provide a common denominator (the observer's sense data) between the minds of (observed) nonhumans and humans. (Details of the relationship between logical positivism and behaviorism are provided by Smith, 1986, and Zuriff, 1984.)

From the point of view of teleological behaviorism, logical positivism is objective about everything except the most important thing—psychological data. It is true that the observer's interior state is a stage between the behavior of the object of study (animal, vegetable, or mineral) and the behavior of the observer. But, as Chapter 5 (on Aristotle's psychology) demonstrates, even the observer's sensations may be viewed as overt actions. It is still less clearly necessary to conceive of the observer's perceptions, cognitions, or the observer's theory itself, in terms of her interior state. As Wittgenstein (1958) pointed out, what ought to count for the validity of a theory of the mind (be it expressed in mathematics, logic, or ordinary language) is its usefulness as a guide for human behavior—not whether or not it matches in 1 : 1 fashion events in the observer's (or the subject's) brain.

2. The passage from Kantor (1963, pp. 345–346) quoted on page 125 continues as follows: "What the scholastics wrought even the behaviorists have been unable to eradicate, but, indeed, have merely carried on in their own fashion."

3. "Raw feels," unlike "intentional acts," for Ryle, are not pure dispositions, but part disposition and part particular act. It is not quite clear whether Ryle considers the part of a raw feel that is not a disposition (not an abstract behavioral category) to be an overt, observable act (as does Aristotle in the present interpretation) or an internal physiological event. To the extent that Ryle means raw feels to be part *internal,* he would not be a teleological behaviorist.

4. Geach's argument would, however, apply to dispositional statements referring to conceivable efficient causation; a hammer that *could* be used to break a vase is certainly not essentially different from a hammer that *could* be used to drive a nail. But his criticism does not apply to dispositional statements referring to *actual behavior;* a hammer that *is being used* to break a vase is being used differently from one that *is being used* to drive in a nail, even if their momentary positions and velocities in space happen to correspond.

5. Lacey and Rachlin (1978) made a more formal but basically similar argument with respect to Fodor's (1974) criticism of behaviorism.

6. This and the next section are an abbreviation of an article, "Pain and Behavior" (Rachlin, 1985).

7. Of course, by pain as distinct from (current) pain behavior, Wittgenstein *could* have meant a temporally extended pattern of pain behavior, rather than a current internal state.

Selected Bibliography

Ackrill, J. L. (1980). Aristotle on eudaimonia. In A. O. Rorty (Ed.), *Essays on Aristotle's ethics* (pp. 15–34). Berkeley and Los Angeles: University of California Press.

Ainslie, G. (1975). Specious reward: A behavioral theory of impulsiveness and impulse control. *Psychological Bulletin, 82,* 436–496.

_____. (1992). *Picoeconomics: The strategic interaction of successive motivational states within the person.* New York: Cambridge University Press.

Allison, J. (1983). *Behavioral economics.* New York: Praeger.

Anscombe, E., & Geach, P. T. (Eds. & Trans.). (1971). *Descartes' philosophical writings.* Indianapolis, IN: Bobbs-Merrill.

Anderson, J. R. (1991). More on rational analysis. *Behavioral and Brain Sciences, 14,* 508–513.

Augustine (c. 395 / 1953). On free will. In J. H. S. Burleigh (Ed. & Trans.), *Augustine: Earlier writings* (pp. 113–218). Philadelphia: Westminster.

Barnes, J. (Ed.). (1984). *The complete works of Aristotle.* 2 volumes. Revised by W. D. Ross. Princeton, NJ: Princeton University Press.

Baum, W. M. (1973). The correlation based law of effect. *Journal of the Experimental Analysis of Behavior, 20,* 137–153.

Becker, G. S. (1976). *The economic approach to human behavior.* Chicago: University of Chicago Press.

Beecher, H. K. (1956). Relationship of significance of wound to the pain experienced. *Journal of the American Medical Association, 161,* 1609.

Block, N. (1981). Psychologism and behaviorism. *Philosophical Review, 90,* 5–43.

Boring, E. G. (1950). *A History of Experimental Psychology.* 2nd ed. New York: Appleton-Century-Crofts.

Breland, K., & Breland, M. (1961). The misbehavior of organisms. *American Psychologist, 16,* 681–684.

Brentano, F. (1973). *Psychology from an empirical standpoint.* L. L. McAlister (Ed.). New York: Humanities Press. (Originally published, 1874. O. Kraus [Ed.], English ed.)

Brunswik, E. (1952). The conceptual framework of psychology. In *International Encyclopedia of Unified Science,* Vol. 1, no. 10. Chicago: University of Chicago Press.

Burleigh, J. H. S. (1953). Prefatory remarks. In his *Augustine: Earlier writings* (pp. 106–107). Philadelphia: Westminster.

Calvino, I. (1988). *Six memos for the next millennium.* Cambridge, MA: Harvard University Press.

Castro, L., & Rachlin, H. (1980). Self-reward, self-monitoring, and self-punishment as feedback in weight control. *Behavior Therapy, 11,* 38–48.

Catania, A. C. (1975). The myth of self-reinforcement. *Behaviorism, 3,* 192–199.

Chomsky, N. (1966). *Cartesian linguistics.* New York: Harper and Row.

————. (1980). Rules and representations. *Behavioral and Brain Sciences, 3,* 1–62.

Danto, A. C. (1983). Science as an intentional system. Comments on an article by D. Dennett. *Behavioral and Brain Sciences, 6,* 359.

Davidson, D. (1980). *Essays on actions and events.* Oxford: Clarendon.

Day, W. (1969). Radical behaviorism in reconciliation with phenomenology. *Journal of the Experimental Analysis of Behavior, 12,* 315–328.

Dennett, D. (1978). *Brainstorms: Philosophical Essays on Mind and Psychology.* Montgomery, VT: Bradford Books.

Descartes (1637 / 1971). Discourse on method (pp. 5–58). In E. Anscombe and P. T. Geach (Eds. & Trans.), *Descartes' philosophical writings.* Indianapolis, IN: Bobbs-Merrill.

————. (1641 / 1971). Meditations (pp. 59–124). In E. Anscombe and P. T. Geach (Eds. & Trans.), *Descartes' philosophical writings.* Indianapolis, IN: Bobbs-Merrill.

Dewey, J. (1896). The reflex arc concept in psychology. *Psychological Review, 3,* 357–370.

Donagan, A. (1987). *Choice: The essential element in human action.* London: Routledge and Kegan Paul.

Dretske, F. (1988). *Explaining behavior: Reasons in a world of causes.* Cambridge, MA: MIT Press.

Ericsson, A. K., & Simon, H. A. (1984). *Protocol analysis: Verbal reports as data.* Cambridge, MA: MIT Press.

Evans, B. (1968). (Ed.). *Dictionary of quotations.* New York: Delacorte.

Fancher, R. E. (1990). *Pioneers of psychology.* 2nd ed. New York: Norton.

Flaubert, G. (1950). *Madame Bovary.* Translated by Alan Russell. New York: Penguin. (Originally published 1856–1857)

Fodor, J. A. (1975). *The language of thought.* New York: Crowell.

————. (1981). *Representations: Philosophical essays on the foundations of cognitive science.* Cambridge, MA: MIT Press.

Fordyce, W. E. (1988). Pain and suffering: A reappraisal. *American Psychologist, 43,* 276–283.

Fordyce, W. E., Fowler, R. S., Lehman, J. F., DeLateur, B. J., Sand, P. L., & Treischmann, R. B. (1974). Operant conditioning in the treatment of chronic pain. In C. M. Franks and G. T. Wilson (Eds.), *Annual review of behavior therapy* (Vol. 2, pp. 691–712). Elmsford, NY: Pergamon.

Friedlander, P. (1958, 1964, 1969). *Plato*. 3 volumes. Princeton, NJ: Princeton University Press.

Gadamer, H-G. (1986). *The idea of the good in Platonic-Aristotelian philosophy.* New Haven, CT: Yale University Press.

Geach, P. T. (1957). *Mental acts.* New York: Humanities Press.

Gibbon, J. (1977). Scalar expectancy theory and Weber's law in animal timing. *Psychological Review, 84,* 279–325.

Green, L., & Rachlin, H. (1991). Economic substitutability of electrical brain stimulation, food, and water. *Journal of the Experimental Analysis of Behavior, 55,* 133–143.

Grossberg, S. (1982). *Studies of mind and brain: Neural principles of learning, perception, development, cognition, and motor control.* Amsterdam: Reidel.

Guthrie, E. R. (1935). *The psychology of learning.* New York: Harper.

Hamilton, E. (1961). Introduction and prefatory notes. In E. Hamilton and H. Cairns (Eds.), *Plato: The collected dialogs* (pp. 575–576). Princeton, NJ: Princeton University Press.

Hayes, S. C. (Ed.). (1989). *Rule-governed behavior: Cognition, contingencies, and instructional control.* New York: Plenum.

Heinrich, B. (1978). *Bumblebee Economics.* Cambridge, MA: Harvard University Press.

Herrnstein, R. J. (1961). Relative and absolute strength of response as a function of frequency of reinforcement. *Journal of the Experimental Analysis of Behavior, 4,* 267–272.

————. (1969). Method and theory in the study of avoidance. *Psychological Review, 76,* 49–69.

————. (1970). On the law of effect. *Journal of the Experimental Analysis of Behavior, 13,* 243–266.

Herrnstein, R. J., & Boring, E. G. (1965). *A source book in the history of psychology.* Cambridge, MA: Harvard University Press.

Herrnstein, R. J., & Vaughan, W., Jr. (1980). Melioration and behavioral allocation. In J. E. R. Staddon (Ed.), *Limits to action: The allocation of individual behavior* (pp. 143–176). New York: Academic Press.

Hilgard, E. R., & Hilgard, J. R. (1975). *Hypnosis in the relief of pain.* Los Altos, CA: William Kaufman.

Hocutt, M. (1974). Aristotle's four becauses. *Philosophy, 49,* 385–399.

Holt, E. B. (1915). *The Freudian wish and its place in ethics.* New York: Henry Holt.

Homme, L. E. (1965). Perspectives in psychology XXIV. Control of coverants, the operants of the mind. *Psychological Record, 15,* 501–511.

Honig, W. K., & Staddon, J. R. (Eds.). (1977). *Handbook of operant behavior.* New York: Prentice-Hall.

Hull, C. L. (1943). *Principles of behavior.* New York: Appleton-Century.

————. (1952). *A behavior system: An introduction to behavior theory concerning the individual organism.* New Haven, CT: Yale University Press.

Hursh, S. R. (1978). The economics of daily consumption controlling food and water-reinforced responding. *Journal of the Experimental Analysis of Behavior, 29,* 475–491.

Huxley, A. (1946). *Brave new world.* New York: Harper and Row.

Irwin, T. N. (1980). The metaphysical and psychological basis of Aristotle's ethics.

In A. O. Rorty (Ed.), *Essays on Aristotle's ethics* (pp. 35–53). Berkeley and Los Angeles: University of California Press.

Kahneman, D., & Tversky, A. (1979). Prospect theory: An analysis of decisions under risk. *Econometrica, 47,* 263–291.

Kantor, J. R. (1963, 1969). *The scientific evolution of psychology.* 2 volumes. Chicago: Principia.

Killeen, P. (1992). Mechanics of the animate. *Journal of the Experimental Analysis of Behavior, 57,* 429–463.

Kimble, G. A. (1961). *Hilgard and Marquis' conditioning and learning.* New York: Appleton-Century-Crofts.

Kuhn, T. S. (1987). What are scientific revolutions? In L. Kruger, L. J. Daston, and M. Heidelberger (Eds.), *The probabilistic revolution: Volume I. Ideas in history* (pp. 7–22). Cambridge, MA: MIT Press.

Lacey, H. M., & Rachlin, H. (1978). Behavior, cognition and theories of choice. *Behaviorism, 6,* 177–202.

Lea, S. E. G., Tarpy, R. M., & Webley, P. (1987). *The individual in the economy.* Cambridge: Cambridge University Press.

Loeb, J. (1918). *Forced movements, tropisms and animal conduct.* Philadelphia: Lippincott.

Loftus, G. (1985). Johannes Kepler's computer simulation of the universe: Some remarks about theory in psychology. *Behavior Research Methods, Instruments, and Computers, 17,* 149–156.

Logue, A. W. (1988). Research on self-control: An integrating framework. *Behavioral and Brain Sciences, 11,* 665–679.

Lucas, J. R. (1970). *The concept of probability.* Oxford: Clarendon.

Machina, M. (1987). Decision-making in the presence of risk. *Science, 236,* 537–542.

Mahoney, M. (1974). *Cognitive behavior modification.* Cambridge, MA: Ballinger.

McGlashen, T. H., Evans, F. J., & Orne, M. T. (1969). The nature of hypnotic analgesia and placebo response to experimental pain. *Psychosomatic Medicine, 31,* 227–246.

McKeon, Richard. (Ed.). (1941). *The basic works of Aristotle.* New York: Random House.

Melzack, R. (1973). How acupuncture can block pain. *Impact of Science on Society, 23,* 1–8.

Merskey, H. (1968). Psychological aspects of pain. *Postgraduate Medical Journal, 44,* 297–306.

Miller, G. (1962). *Psychology: The science of mental life.* Harmondsworth, Middlesex, England: Penguin.

Mischel, W., Shoda, Y., & Rodriguez, M. (1989). Delay of gratification in children. *Science, 244,* 933–938.

Nagel, T. (1974). What is it like to be a bat? *Philosophical Review, 83,* 435–450.

——————. (1980). Aristotle on *eudaimonia.* In A. O. Rorty (Ed.), *Essays on Aristotle's ethics* (pp. 7–14). Berkeley and Los Angeles: University of California Press.

Nisbett, R. E., & Wilson, T. D. (1977). Telling more than we can know: Verbal reports on mental processes. *Psychological Review, 84,* 231–259.

Nussbaum, M. (1980). Shame, separateness and political unity: Aristotle's criticism of Plato. In A. O. Rorty (Ed.), *Essays on Aristotle's ethics* (pp. 395–436). Berkeley and Los Angeles: University of California Press.

Pavlov, I. P. (1927). *Conditioned reflexes.* Translated by G. V. Anrep. London: Oxford University Press.

Plato (1961). *The collected dialogs.* Edited by E. Hamilton & H. Cairnes. Princeton, NJ: Princeton University Press.

Premack, D. (1965). Reinforcement theory. In D. Levine (Ed.), *Nebraska symposium on motivation: 1965* (pp. 123–179). Lincoln: University of Nebraska Press.

Putnam, H. (1980). Brains and behavior. In N. Block (Ed.), *Readings in philosophy of psychology* (Vol. 1, pp. 24–36). Cambridge, MA: Harvard University Press.

Quine, W. V. O. (1960). *Word and object.* Cambridge, MA: MIT Press.

Rachlin, H. (1978). A molar theory of reinforcement schedules. *Journal of the Experimental Analysis of Behavior, 30,* 345–360.

_____. (1980). *Behaviorism in everyday life.* Englewood Cliffs, NJ: Prentice-Hall.

_____. (1985). Pain and behavior. *Behavioral and Brain Sciences, 8,* 43–83.

_____. (1989). *Judgment, decision and choice.* New York: Freeman.

_____. (1991). *Introduction to modern behaviorism.* 3rd ed. New York: Freeman.

_____. (1992). Teleological behaviorism. *American Psychologist, 47,* 1371–1382.

Rachlin, H., Battalio, R., Kagel, J., & Green, L. (1981). Maximization theory in behavioral psychology. *Behavioral and Brain Sciences, 4,* 371–388.

Rachlin, H., & Burkhard, B. (1978). The temporal triangle: Response substitution in instrumental conditioning. *Psychological Review, 85,* 22–48.

Rachlin, H., & Green, L. (1972). Commitment, choice and self-control. *Journal of the Experimental Analysis of Behavior, 17,* 15–22.

Rachlin, H., Logue, A. W., Gibbon, J., & Frankel, M. (1986). Cognition and behavior in studies of choice. *Psychological Review, 93,* 33–45.

Randall, J. H., Jr. (1960). *Aristotle.* New York: Columbia University Press.

Rescorla, R. A. (1988). Pavlovian conditioning: It's not what you think it is. *American Psychologist, 43,* 151–160.

Ringen, J. (1985). Operant conditioning and a paradox of teleology. *Philosophy of Science, 52,* 565–577.

Robinson, D. N. (1989). *Aristotle's psychology.* New York: Columbia University Press.

Rorty, R. (1979). *Philosophy and the mirror of nature.* Princeton, NJ: Princeton University Press.

_____. (1982). *Consequences of pragmatism: Essays, 1972–1980.* Minneapolis: University of Minnesota Press.

Rychalk, J. (1988). *The psychology of rigorous humanism.* 2nd ed. New York: New York University Press.

Ryle, G. (1949). *The concept of mind.* London: Hutchinson House.

Samuelson, P. A. (1973). *Economics: An introductory analysis.* 9th ed. New York: McGraw-Hill.

Savin, H. (1980). Introduction: Behaviorism. In N. Block (Ed.), *Readings in philosophy of psychology* (pp. 11–13). Cambridge, MA: Harvard University Press.

Schwartz, B., & Lacey, H. (1982). *Behaviorism, science and human nature.* New York: Norton.

Searle, J. R. (1980). Minds, brains and programs. *Behavioral and Brain Sciences, 3,* 324.

_____. (1981). Analytic philosophy and mental phenomena. In P. A. French, T. E. Vehgling, Jr., & H. K. Wettstein (Eds.), *Midwest studies in philosophy VI* (pp. 405–423). Minneapolis: University of Minnesota Press.

_____. (1989). Reply to: Artificial intelligence and the Chinese room by E. Motzkin. *New York Review of Books,* 36, no. 2, p. 45.

Sechenov, I. M. (1965). Reflexes of the brain. In R. J. Herrnstein & E. G. Boring (Eds.), *A source book in the history of psychology* (pp. 308–320). Cambridge, MA: Harvard University Press. (Originally published 1863)

Shepard, R. N., & Metzler, J. (1971). Mental rotation of three-dimensional objects. *Science, 171,* 701–703.

Sherrington, C. S. (1906). *The integrative action of the nervous system.* New York: Scribner.

Silverstein, A. (1988). An Aristotelian resolution of the ideographic versus nomothetic tension. *American Psychologist, 43,* 425–430.

Skinner, B. F. (1938). *The behavior of organisms.* New York: Appleton-Century.

_____. (1948). *Walden Two.* New York: Macmillan.

_____. (1953). *Science and human behavior.* New York: Macmillan.

_____. (1957). *Verbal behavior.* New York: Appleton-Century-Crofts.

_____. (1974). *About behaviorism.* New York: Knopf.

_____. (1979). *The shaping of a behaviorist.* New York: Knopf.

Small, W. S. (1901). Experimental study of the mental processes of the rat. *American Journal of Psychology, 12,* 218–220.

Smith, L. D. (1986). *Behaviorism and logical positivism: A reassessment of the alliance.* Stanford, CA: Stanford University Press.

Staddon, J. E. R. (1973). On the notion of cause with applications to behaviorism. *Behaviorism, 1,* 25–64.

_____. (1979). Operant behavior as adaptation to constraint. *Journal of Experimental Psychology: General, 108,* 48–67.

Staddon, J. E. R., & Motheral, S. (1978). On matching and maximizing in operant choice experiments. *Psychological Review, 85,* 436–445.

Stevens, S. S. (1936). Psychology: The propaedeutic science. *Philosophy of Science, 3,* 90–103.

Stich, S. (1983). *From folk psychology to cognitive science.* Cambridge, MA: MIT Press.

Taylor, C. (1964). *The explanation of behavior.* London: Routledge and Kegan Paul.

Teitelbaum, P. (1977). Levels of integration of the operant. In J. E. R. Staddon and W. K. Honig (Eds.), *Handbook of operant behavior* (pp. 7–27). Englewood Cliffs, NJ: Prentice-Hall.

Thorndike, E. L. (1911). *Animal intelligence.* New York: Macmillan.

Timberlake, W., & Allison, J. (1974). Response deprivation: An empirical approach to instrumental performance. *Psychological Review, 81,* 146–164.

Tolman, E. C. (1932). *Purposive behavior in animals and men.* New York: Appleton-Century.

_____. (1949). There is more than one kind of learning. *Psychological Review, 56,* 144–155.

Turk, D. C., Meichenbaum, D., & Genest, M. (1983). *Pain and behavioral medicine: A cognitive-behavioral perspective.* New York: Guilford Press.

Tursky, B., & Sternbach, R. A. (1967). Further physiological correlates of ethnic differences in responses to shock. *Psychophysiology, 4,* 67–74.

Vlastos, G. (1975). *Plato's universe*. Seattle: University of Washington Press.

Watkins, M. J. (1990). Mediationism and the obfuscation of memory. *American Psychologist, 45,* 328–335.

Watson, J. B. (1913). Psychology as the behaviorist views it. *Psychological Review, 20,* 158–177.

Watson, R. I., & Evans, R. B. (1991). *The great psychologists: A history of psychological thought.* 5th ed. New York: Harper Collins.

Wittgenstein, L. (1958). *Philosophical Investigations.* Translated by G. E. M. Anscombe. New York: Macmillan.

Wright, L. (1976). *Teleological explanations.* Berkeley and Los Angeles: University of California Press.

Yourgrau, W., & Mandelstam, S. (1968). *Variational principles in dynamics and quantum theory.* New York: Dover.

Zuriff, G. E. (1979). Ten inner causes. *Behaviorism, 7,* 1–8.

———. (1984). *Behaviorism: A conceptual reconstruction.* New York: Columbia University Press.

Index

Helmholtz, H. von, 5
Herrnstein, R. J., *ix,* 19*n*.7, 32, 39*n*.6
Hilgard, E. R., 147
Hilgard, J. R., 147
Hocutt, M., 20, 24
Hofmansthal, H. von, 88
Holism. *See* Gestalt psychology; Molarism
Holt, E. B., 18*n*.5
Homme, L. E., 30
Honig, W. K., 29
Hull, C. L., 3, 9–11, 12, 24–27
Hursh, S. R., 148
Huxley, A., 68

Imagination, 21, 40*n*.9, 98–102, 129
  and thought, 100–102
Information, 33
Information-processing systems, 138
Intelligence, 137–42
Intentional acts, 82, 113*n*.4, 150*n*.3
Intentional states, 136, 137–42, 146
Interbehaviorism, 42, 61
Introspection, 4–6, 37–39, 60, 118–22,
  124, 128–29, 145–46
Irwin, T. N., 23, 24, 88, 108
Isomorphism, 7

James, W., 6
Jennings, H. S., 11–12, 18*n*.4
Justice, 56, 65*n*.5, 106, 113*n*.7

Kagel, J., 149
Kahneman, D., 35
Kant, I., 18*n*.2, 132
Kantor, J. R., *vi,* 38, 41–42, 46, 61, 67,
  115, 121, 123, 125, 150*n*.2
Killeen, P., 40*n*.6
Kimble, G. A., 9
Knowledge, 9, 55–62, 68–70, 73, 77, 88,
  90, 96, 111, 130–31, 139, 149
Koffka, K., 7
Köhler, W., 7
Kuhn, T., 80–84, 85*n*.7

Lacey, H., *ix,* 50, 150*n*.5
Language, 62–63. *See also* Verbal behavior;
  Verbal reports
  and the control of behavior, 108–10
Lea, S. E. G., 148
Learning, 60–62, 73
Lehman, J. F., 148
Lewin, K., 7
Lloyd Morgan, C., 10
Locke, J., 5–6
Loeb, J., 11–12
Loftus, G., *viii*

Logic, 72–74, 75, 77, 81, 132
  and ethics, 113*n*.9
Logical positivism, 150*n*.1
Logue, A. W., 36
Lotze, R. H., 112*n*.2
Love, 16–18, 53–55
Lucas, J. R., 35

Machina, M., 64*n*.4
Mahoney, M., 30
Manichaeanism, 117
Material causes. *See* Causation, material
McGlashen, T. H., 147
Meaning, 109
Mechanisms, 28, 35–36
  mental, *v–viii*
Mediation, cognitive, 25–26
Melzack, R., 147
Mencken, H. L., 105
Mental chemistry, 5
Mental life, *v*
Mental states, 89, 136
Mental terms, 15–16, 28, 38–39, 146
Mentalism, 15, 34
Merskey, H., 147
Metzler, J., 5
Mill, J. S., 5, 132
Miller, G., *vi*
Mischel, W., 100–101, 109–10
Molarism, 38, 53, 61, 65*n*.6, 148
Moral behavior. *See* Ethics
Motheral, S., 53
Motivation, 33–36
  elite of, 122
Movement, 31, 80–84, 87–88, 90, 92,
  94–95, 96–98, 117
Müller, J., 4–6
Myth, in Plato, 43, 57

Nagel, T., 103–4, 112*n*.2
Naturalism, *vi,* 41–42, 67, 116
Neoplatonism, 46, 117–23, 132*n*.2
Neorealism, 18*n*.5
Nisbett, R. E., 143
Nussbaum, M., 84*n*.2

Opacity, 32, 132*n*.3
Operant conditioning, 14, 27, 29–33,
  60–62, 79–80, 84
Orne, M. T., 147
Ovid, 105

Pain, 142–48
Pavlov, I. P., 9, 24, 32
Perception, 21, 50, 72–74, 96–98
Pflüger, E. F. W., 112*n*.2

DATE DUE